generatingfunctionology

second edition

generatingfunctionology

second edition

Herbert S. Wilf
Department of Mathematics
University of Pennsylvania
Philadelphia, Pennsylvania

Academic Press, Inc.
Harcourt Brace & Company, Publishers
Boston   San Diego   New York
London   Sydney   Tokyo   Toronto

ACADEMIC PRESS, INC.
525 B Street, Suite 1900, San Diego, California 92101-4495

United Kingdom Edition published by
ACADEMIC PRESS LIMITED
24–28 Oval Road, London NW1 7DX

ISBN 0-12-751956-4

Printed in the United States of America
93 94 95 96 97 BB 9 8 7 6 5 4 3 2 1

CONTENTS

Chapter 1: Introductory Ideas and Examples

Chapter 2: Series

Chapter 3: Cards, Decks, and Hands: The Exponential Formula

Chapter 4: Applications of generating functions

Chapter 5: Analytic and asymptotic methods

Preface

This book is about generating functions and some of their uses in discrete mathematics. The subject is so vast that I have not attempted to give a comprehensive discussion. Instead I have tried only to communicate some of the main ideas.

Generating functions are a bridge between discrete mathematics, on the one hand, and continuous analysis (particularly complex variable theory) on the other. It is possible to study them solely as tools for solving discrete problems. As such there is much that is powerful and magical in the way generating functions give unified methods for handling such problems. The reader who wished to omit the analytical parts of the subject would skip chapter 5 and portions of the earlier material.

To omit those parts of the subject, however, is like listening to a stereo broadcast of, say, Beethoven's Ninth Symphony, using only the left audio channel.

The full beauty of the subject of generating functions emerges only from tuning in on both channels: the discrete and the continuous. See how they make the solution of difference equations into child's play. Then see how the theory of functions of a complex variable gives, virtually by inspection, the approximate size of the solution. The interplay between the two channels is vitally important for the appreciation of the music.

In recent years there has been a vigorous trend in the direction of finding bijective proofs of combinatorial theorems. That is, if we want to prove that two sets have the same cardinality then we should be able to do it by exhibiting an explicit bijection between the sets. In many cases the fact that the two sets have the same cardinality was discovered in the first place by generating function arguments. Also, even though bijective arguments may be known, the generating function proofs may be shorter or more elegant.

The bijective proofs give one a certain satisfying feeling that one 'really' understands why the theorem is true. The generating function arguments often give satisfying feelings of naturalness, and 'oh, I could have thought of that,' as well as usually offering the best route to finding exact or approximate formulas for the numbers in question.

This book was tested in a senior course in discrete mathematics at the University of Pennsylvania. My thanks go to the students in that course for helping me at least partially to debug the manuscript, and to a number of my colleagues who have made many helpful suggestions. Any reader who is kind enough to send me a correction will receive a then-current complete errata sheet and many thanks.

Herbert S. Wilf
Philadelphia, PA
September 1, 1989

Preface to the Second Edition

This edition contains several new areas of application, in chapter 4, many new problems and solutions, a number of improvements in the presentation, and corrections. It also contains an Appendix that describes some of the features of computer algebra programs that are of particular importance in the study of generating functions.

I am indebted to many people for helping to make this a better book. Bruce Sagan, in particular, made many helpful suggestions as a result of a test run in his classroom. Many readers took up my offer (which is now repeated) to supply a current errata sheet and my thanks in return for any errors discovered.

Herbert S. Wilf
Philadelphia, PA
May 21, 1992

Chapter 1
Introductory ideas and examples

A generating function is a clothesline on which we hang up a sequence of numbers for display.

What that means is this: suppose we have a problem whose answer is a sequence of numbers, $a_0, a_1, a_2, \ldots$. We want to 'know' what the sequence is. What kind of an answer might we expect?

A simple formula for a_n would be the best that could be hoped for. If we find that $a_n = n^2 + 3$ for each $n = 0, 1, 2, \ldots$, then there's no doubt that we have 'answered' the question.

But what if there isn't any simple formula for the members of the unknown sequence? After all, some sequences are complicated. To take just one hair-raising example, suppose the unknown sequence is 2, 3, 5, 7, 11, 13, 17, 19, $\ldots$, where a_n is the nth prime number. Well then, it would be just plain unreasonable to expect any kind of a simple formula.

Generating functions add another string to your bow. Although giving a simple formula for the members of the sequence may be out of the question, we might be able to give a simple formula for *the sum of a power series, whose coefficients are the sequence that we're looking for.*

For instance, suppose we want the Fibonacci numbers $F_0, F_1, F_2, \ldots$, and what we know about them is that they satisfy the recurrence relation

$$F_{n+1} = F_n + F_{n-1} \qquad (n \geq 1; F_0 = 0; F_1 = 1).$$

The sequence begins with 0, 1, 1, 2, 3, 5, 8, 13, 21, 34, 55, $\ldots$ There *are* exact, not-very-complicated formulas for F_n, as we will see later, in example 2 of this chapter. But, just to get across the idea of a generating function, here is how a generatingfunctionologist might answer the question: *the nth Fibonacci number, F_n, is the coefficient of x^n in the expansion of the function $x/(1 - x - x^2)$ as a power series about the origin.*

You may concede that this is a kind of answer, but it leaves a certain unsatisfied feeling. It isn't *really* an answer, you might say, because we don't have that explicit formula. Is it a *good* answer?

In this book we hope to convince you that answers like this one are often spectacularly good, in that they are themselves elegant, they allow you to do almost anything you'd like to do with your sequence, and generating functions can be simple and easy to handle even in cases where exact formulas might be stupendously complicated.

Here are some of the things that you'll often be able to do with generating function answers:

(a) **Find an exact formula for the members of your sequence.**
Not always. Not always in a pleasant way, if your sequence is

complicated. But at least you'll have a good shot at finding such a formula.

(b) **Find a recurrence formula.** Most often generating functions arise from recurrence formulas. Sometimes, however, from the generating function you will find a new recurrence formula, not the one you started with, that gives new insights into the nature of your sequence.

(c) **Find averages and other statistical properties of your sequence.** Generating functions can give stunningly quick derivations of various probabilistic aspects of the problem that is represented by your unknown sequence.

(d) **Find asymptotic formulas for your sequence.** Some of the deepest and most powerful applications of the theory lie here. Typically, one is dealing with a very difficult sequence, and instead of looking for an *exact* formula, which might be out of the question, we look for an *approximate* formula. While we would not expect, for example, to find an exact formula for the nth prime number, it is a beautiful fact (the 'Prime Number Theorem') that the nth prime is *approximately* $n \log n$ when n is large, in a certain precise sense. In chapter 5 we will discuss asymptotic problems.

(e) **Prove unimodality, convexity, etc.** A sequence is called *unimodal* if it increases steadily at first, and then decreases steadily. Many combinatorial sequences are unimodal, and a variety of methods are available for proving such theorems. Generating functions can help. There are methods by which the analytic properties of the generating function can be translated into conclusions about the rises and falls of the sequence of coefficients. When the method of generating functions works, it is often the simplest method known.

(f) **Prove identities.** Many, many identities are known, in combinatorics and elsewhere in mathematics. The identities that we refer to are those in which a certain formula is asserted to be equal to another formula for stated values of the free variable(s). For example, it is well known that

$$\sum_{j=0}^{n} \binom{n}{j}^2 = \binom{2n}{n} \qquad (n = 0, 1, 2, \ldots).$$

One way to prove such identities is to consider the generating function whose coefficients are the sequence shown on the left side of the claimed identity, and to consider the generating function formed from the sequence on the right side of the claimed identity, and to show that these are the same function. This may sound obvious, but it is quite remarkable how much simpler and more transparent many of the derivations become when seen from the point of view of the black belt generatingfunctionologist. The 'Snake Oil' method that we present in section 4.3, below, explores some of these vistas. The method of rational functions, in section 4.4, is new, and does more and harder problems of this kind.

(g) **Other.** Is there something else you would like to know about your sequence? A generating function may offer hope. One example might be the discovery of congruence relations. Another possibility is that your generating function may bear a striking resemblance to some other known generating function, and that may lead you to the discovery that your problem is closely related to another one, which you never suspected before. It is noteworthy that in this way you may find out that the answer to your problem is simply related to the answer to another problem, without knowing formulas for the answers to either one of the problems!

In the rest of this chapter we are going to give a number of examples of problems that can be profitably thought about from the point of view of generating functions. We hope that after studying these examples the reader will be at least partly convinced of the power of the method, as well as of the beauty of the unified approach.

1.1 An easy two term recurrence

A certain sequence of numbers $a_0, a_1, \ldots$ satisfies the conditions

$$a_{n+1} = 2a_n + 1 \qquad (n \geq 0; a_0 = 0). \qquad (1.1.1)$$

Find the sequence.

First try computing a few members of the sequence to see what they look like. It begins with 0, 1, 3, 7, 15, 31, ... These numbers look suspiciously like 1 less than the powers of 2. So we could conjecture that $a_n = 2^n - 1$ $(n \geq 0)$, and prove it quickly, by induction based on the recurrence (1.1.1).

But this is a book about generating functions, so let's forget all of that, pretend we didn't spot the formula, and use the generating function method. Hence, instead of finding the sequence $\{a_n\}$, let's find the generating function $A(x) = \sum_{n \geq 0} a_n x^n$. Once we know what that function is, we will be able to read off the explicit formula for the a_n's by expanding $A(x)$ in a series.

To find $A(x)$, multiply both sides of the recurrence relation (1.1.1) by x^n and sum over the values of n for which the recurrence is valid, namely, over $n \geq 0$. Then try to relate these sums to the unknown generating function $A(x)$.

If we do this first to the left side of (1.1.1), there results $\sum_{n\geq 0} a_{n+1} x^n$. How can we relate this to $A(x)$? It is *almost* the same as $A(x)$. But the subscript of the 'a' in each term is 1 unit larger than the power of x. But, clearly,

$$\sum_{n\geq 0} a_{n+1} x^n = a_1 + a_2 x + a_3 x^2 + a_4 x^3 + \cdots$$

$$= \{(a_0 + a_1 x + a_2 x^2 + a_3 x^3 + \cdots) - a_0\}/x$$
$$= A(x)/x$$

since $a_0 = 0$ in this problem. Hence the result of so operating on the left side of (1.1.1) is $A(x)/x$.

Next do the right side of (1.1.1). Multiply it by x^n and sum over all $n \geq 0$. The result is

$$\sum_{n\geq 0}(2a_n + 1)x^n = 2A(x) + \sum_{n\geq 0} x^n$$

$$= 2A(x) + \frac{1}{1-x},$$

wherein we have used the familiar geometric series evaluation $\sum_{n\geq 0} x^n = 1/(1-x)$, which is valid for $|x| < 1$.

If we equate the results of operating on the two sides of (1.1.1), we find that

$$\frac{A(x)}{x} = 2A(x) + \frac{1}{1-x},$$

which is trivial to solve for the unknown generating function $A(x)$, in the form

$$A(x) = \frac{x}{(1-x)(1-2x)}.$$

This is the generating function for the problem. The unknown numbers a_n are arranged neatly on this clothesline: a_n is the coefficient of x^n in the series expansion of the above $A(x)$.

Suppose we want to find an explicit formula for the a_n's. Then we would have to expand $A(x)$ in a series. That isn't hard in this example, since the partial fraction expansion is

$$\frac{x}{(1-x)(1-2x)} = x\left\{\frac{2}{1-2x} - \frac{1}{1-x}\right\}$$
$$= \{2x + 2^2 x^2 + 2^3 x^3 + 2^4 x^4 + \cdots\}$$
$$- \{x + x^2 + x^3 + x^4 + \cdots\}$$
$$= (2-1)x + (2^2 - 1)x^2 + (2^3 - 1)x^3 + (2^4 - 1)x^4 + \cdots$$

It is now clear that the coefficient of x^n, i.e. a_n, is equal to $2^n - 1$, for each $n \geq 0$.

In this example, the heavy machinery wasn't needed because we knew the answer almost immediately, by inspection. The impressive thing about generatingfunctionology is that even though the problems can get a lot harder than this one, the method stays very much the same as it was here, so the same heavy machinery may produce answers in cases where answers are not a bit obvious.

1.2 A slightly harder two term recurrence

A certain sequence of numbers $a_0, a_1, \ldots$ satisfies the conditions

$$a_{n+1} = 2a_n + n \qquad (n \geq 0; a_0 = 1). \qquad (1.2.1)$$

Find the sequence.

As before, we might calculate the first several members of the sequence, to get 1, 2, 5, 12, 27, 58, 121, ... A general formula does not seem to be immediately in evidence in this case, so we use the method of generating functions. That means that instead of looking for the *sequence* $a_0, a_1, \ldots$, we will look for the *function* $A(x) = \sum_{j \geq 0} a_j x^j$. Once we have found the function, the sequence will be identifiable as the sequence of power series coefficients of the function.*

As in example 1, the first step is to make sure that the recurrence relation that we are trying to solve comes equipped with a clear indication of the range of values of the subscript for which it is valid. In this case, the recurrence (1.2.1) is clearly labeled in the parenthetical comment as being valid for $n = 0, 1, 2, \ldots$ Don't settle for a recurrence that has an unqualified free variable.

The next step is to define the generating function that you will look for. In this case, since we are looking for a sequence $a_0, a_1, a_2, \ldots$ one natural choice would be the function $A(x) = \sum_{j \geq 0} a_j x^j$ that we mentioned above.

Next, take the recurrence relation (1.2.1), multiply both sides of it by x^n, and sum over *all the values of n for which the relation is valid,* which, in this case, means sum from $n = 0$ to ∞. Try to express the result of doing that in terms of the function $A(x)$ that you have just defined.

If we do that to the left side of (1.2.1), the result is

$$a_1 + a_2 x + a_3 x^2 + a_4 x^3 + \cdots = (A(x) - a_0)/x$$
$$= (A(x) - 1)/x.$$

So much for the left side. What happens if we multiply the right side of (1.2.1) by x^n and sum over nonnegative integers n? Evidently the result is $2A(x) + \sum_{n \geq 0} n x^n$. We need to identify the series

$$\sum_{n \geq 0} n x^n = x + 2x^2 + 3x^3 + 4x^4 + \cdots$$

---

* If you are feeling rusty in the power series department, see chapter 2, which contains a review of that subject.

There are two ways to proceed: (a) look it up (b) work it out. To work it out we use the following stunt, which seems artificial if you haven't seen it before, but after using it 4993 times it will seem quite routine:

$$\sum_{n\geq 0} nx^n = \sum_{n\geq 0} x(\frac{d}{dx})x^n = x(\frac{d}{dx})\sum_{n\geq 0} x^n = x(\frac{d}{dx})\frac{1}{1-x} = \frac{x}{(1-x)^2}.$$

(1.2.2)

In other words, the series that we are interested in is essentially the derivative of the geometric series, so its sum is essentially the derivative of the sum of the geometric series. This raises some nettlesome questions, which we will mention here and deal with later. For what values of x is (1.2.2) valid? The geometric series converges only for $|x| < 1$, so the analytic manipulation of functions in (1.2.2) is legal only for those x. However, often the *analytic* nature of the generating function doesn't interest us; we love it only for its role as a clothesline on which our sequence is hanging out to dry. In such cases we can think of a generating function as only a *formal* power series, i.e., as an algebraic object rather than as an analytic one. Then (1.2.2) would be valid as an identity in the ring of formal power series, which we will discuss later, and the variable x wouldn't need to be qualified at all.

Anyway, the result of multiplying the right hand side of (1.2.1) by x^n and summing over $n \geq 0$ is $2A(x) + x/(1-x)^2$, and if we equate this with our earlier result from the left side of (1.2.1), we find that

$$\frac{(A(x)-1)}{x} = 2A(x) + \frac{x}{(1-x)^2},$$

(1.2.3)

and we're ready for the easy part, which is to solve (1.2.3) for the unknown $A(x)$, getting

$$A(x) = \frac{1 - 2x + 2x^2}{(1-x)^2(1-2x)}.$$

(1.2.4)

Exactly what have we learned? The original problem was to 'find' the numbers $\{a_n\}$ that are determined by the recurrence (1.2.1). We have, in a certain sense, 'found' them: the number a_n is the coefficient of x^n in the power series expansion of the function (1.2.4).

This is the end of the 'find-the-generating-function' part of the method. We have it. What we do with it depends on exactly why we wanted to know the solution of (1.2.1) in the first place.

Suppose, for example, that we want an exact, simple formula for the members a_n of the unknown sequence. Then the method of partial fractions will work here, just as it did in the first example, but its application is now a little bit trickier. Let's try it and see.

The first step is to expand the right side of (1.2.4) in partial fractions. Such a fraction is guaranteed to be expandable in partial fractions in the form

$$\frac{1 - 2x + 2x^2}{(1 - x)^2(1 - 2x)} = \frac{A}{(1 - x)^2} + \frac{B}{1 - x} + \frac{C}{1 - 2x}, \qquad (1.2.5)$$

and the only problem is how to find the constants A, B, C.

Here's the quick way. First multiply both sides of (1.2.5) by $(1 - x)^2$, and then let $x = 1$. The instant result is that $A = -1$ (don't take my word for it, try it for yourself!). Next multiply (1.2.5) through by $1 - 2x$ and let $x = 1/2$. The instant result is that $C = 2$. The hard one to find is B, so let's do that one by cheating. Since we know that (1.2.5) is an identity, i.e., is true for all values of x, let's choose an easy value of x, say $x = 0$, and substitute that value of x into (1.2.5). Since we now know A and C, we find at once that $B = 0$.

We return now to (1.2.5) and insert the values of A, B, C that we just found. The result is the relation

$$A(x) = \frac{1 - 2x + 2x^2}{(1 - x)^2(1 - 2x)} = \frac{(-1)}{(1 - x)^2} + \frac{2}{1 - 2x}. \qquad (1.2.6)$$

What we are trying to do is to find an explicit formula for the coefficient of x^n in the left side of (1.2.6). We are trading that in for two easier problems, namely finding the coefficient of x^n in each of the summands on the right side of (1.2.6). Why are they easier? The term $2/(1 - 2x)$, for instance, expands as a geometric series. The coefficient of x^n there is just $2 \cdot 2^n = 2^{n+1}$. The series $(-1)/(1 - x)^2$ was handled in (1.2.2) above, and its coefficient of x^n is $-(n + 1)$. If we combine these results we see that our unknown sequence is

$$a_n = 2^{n+1} - n - 1 \qquad (n = 0, 1, 2, \ldots).$$

Having done all of that work, it's time to confess that there are better ways to deal with recurrences of the type (1.2.1), without using generating functions.* However, the problem remains a good example of how generating functions can be used, and it underlines the fact that a single unified method can replace a lot of individual special techniques in problems about sequences. Anyway, it won't be long before we're into some problems that essentially cannot be handled without generating functions. ∎

It's time to introduce some notation that will save a lot of words in the sequel.

Definition. *Let $f(x)$ be a series in powers of x. Then by the symbol $[x^n]f(x)$ we will mean the coefficient of x^n in the series $f(x)$.*

Here are some examples of the use of this notation.

$$[x^n]e^x = 1/n!; \quad [t^r]\{1/(1 - 3t)\} = 3^r; \quad [u^m](1 + u)^s = \binom{s}{m}.$$

---

* See, for instance, chapter 1 of my book [Wil].

A perfectly obvious property of this symbol, that we will use repeatedly, is

$$[x^n]\{x^a f(x)\} = [x^{n-a}]f(x). \tag{1.2.7}$$

Another property of this symbol is the convention that if β is any real number, then

$$[\beta x^n]f(x) = (1/\beta)[x^n]f(x), \tag{1.2.8}$$

so, for instance, $[x^n/n!]e^x = 1$ for all $n \geq 0$.

Before we move on to the next example, here is a summary of the method of generating functions as we have used it so far.

THE METHOD

Given: a recurrence formula that is to be solved by the method of generating functions.

1. Make sure that the set of values of the free variable (say n) for which the given recurrence relation is true, is clearly delineated.
2. Give a name to the generating function that you will look for, and write out that function in terms of the unknown sequence (e.g., call it $A(x)$, and define it to be $\sum_{n\geq 0} a_n x^n$).
3. Multiply both sides of the recurrence by x^n, and sum over all values of n for which the recurrence holds.
4. Express both sides of the resulting equation explicitly in terms of your generating function $A(x)$.
5. Solve the resulting equation for the unknown generating function $A(x)$.
6. If you want an exact formula for the sequence that is defined by the given recurrence relation, then attempt to get such a formula by expanding $A(x)$ into a power series by any method you can think of. In particular, if $A(x)$ is a rational function (quotient of two polynomials), then success will result from expanding in partial fractions and then handling each of the resulting terms separately.

1.3 A three term recurrence

Now let's do the Fibonacci recurrence

$$F_{n+1} = F_n + F_{n-1}. \qquad (n \geq 1; F_0 = 0; F_1 = 1). \tag{1.3.1}$$

Following 'The Method,' we will solve for the generating function

$$F(x) = \sum_{n\geq 0} F_n x^n.$$

To do that, multiply (1.3.1) by x^n, and sum over $n \geq 1$. We find on the left side

$$F_2 x + F_3 x^2 + F_4 x^3 + \cdots = \frac{F(x) - x}{x},$$

and on the right side we find

$$\{F_1 x + F_2 x^2 + F_3 x^3 + \cdots\} + \{F_0 x + F_1 x^2 + F_2 x^3 + \cdots\} = \{F(x)\} + \{x F(x)\}.$$

(Important: Try to do the above yourself, without peeking, and see if you get the same answer.) It follows that $(F - x)/x = F + xF$, and therefore that the unknown generating function is now known, and it is

$$F(x) = \frac{x}{1 - x - x^2}.$$

Now we will find some formulas for the Fibonacci numbers by expanding $x/(1 - x - x^2)$ in partial fractions. The success of the partial fraction method is greatly enhanced by having only linear (first degree) factors in the denominator, whereas what we now have is a quadratic factor. So let's factor it further. We find that

$$1 - x - x^2 = (1 - xr_+)(1 - xr_-) \qquad (r_\pm = (1 \pm \sqrt{5})/2)$$

and so

$$\frac{x}{1 - x - x^2} = \frac{x}{(1 - xr_+)(1 - xr_-)}$$

$$= \frac{1}{(r_+ - r_-)} \left(\frac{1}{1 - xr_+} - \frac{1}{1 - xr_-} \right)$$

$$= \frac{1}{\sqrt{5}} \left\{ \sum_{j \geq 0} r_+^j x^j - \sum_{j \geq 0} r_-^j x^j \right\},$$

thanks to the magic of the geometric series. It is easy to pick out the coefficient of x^n and find

$$F_n = \frac{1}{\sqrt{5}} (r_+^n - r_-^n) \qquad (n = 0, 1, 2, \ldots) \tag{1.3.3}$$

as an explicit formula for the Fibonacci numbers F_n.

This example offers us a chance to edge a little further into what generating functions can tell us about sequences, in that we can get not only the exact answer, but also an approximate answer, valid when n is large. Indeed, when n is large, since $r_+ > 1$ and $|r_-| < 1$, the second term in (1.3.3) will be minuscule compared to the first, so an extremely good approximation to F_n will be

$$F_n \sim \frac{1}{\sqrt{5}} \left(\frac{1 + \sqrt{5}}{2} \right)^n. \tag{1.3.4}$$

But, you may ask, why would anyone want an approximate formula when an exact one is available? One answer, of course, is that sometimes exact answers are fearfully complicated, and approximate ones are more revealing. Even in this case, where the exact answer isn't very complex, we can still learn something from the approximation. The reader should take a few moments to verify that, by neglecting the second term in (1.3.3), we neglect a quantity that is never as large as 0.5 in magnitude, and consequently not only is F_n approximately given by (1.3.4), it is *exactly* equal to the integer nearest to the right side of (1.3.4). Thus consideration of an approximate formula has found us a simpler exact formula!

1.4 A three term boundary value problem

This example will differ from the previous ones in that the recurrence relation involved does not permit the direct calculation of the members of the sequence, although it does determine the sequence uniquely. The situation is similar to the following: suppose we imagine the Fibonacci recurrence, together with the additional data $F_0 = 1$ and $F_{735} = 1$. Well then, the sequence $\{F_n\}$ would be uniquely determined, but you wouldn't be able to compute it directly by recurrence because you would not be in possession of the two consecutive values that are needed to get the recurrence started.

We will consider a slightly more general situation. It consists of the recurrence

$$ay_{n+1} + by_n + cy_{n-1} = d_n \qquad (n = 1, 2, \ldots, N-1; y_0 = y_N = 0) \quad (1.4.1)$$

where the positive integer N, the constants a, b, c and the sequence $\{d_n\}_{n=1}^{N-1}$ are given in advance. The equations (1.4.1) determine the sequence $\{y_i\}_0^N$ uniquely, as we will see, and the method of generating functions gives us a powerful way to attack such boundary value problems as this, which arise in numerous applications, such as the theory of interpolation by spline functions.

To begin with, we will define two generating functions. One of them is our unknown $Y(x) = \sum_{j=0}^{N} y_j x^j$, and the second one is $D(x) = \sum_{j=1}^{N-1} d_j x^j$, and it is regarded as a known function (did we omit any given values of the d_j's, like d_0? or d_N? Why?).

Next, following the usual recipe, we multiply the recurrence (1.4.1) by x^n and sum over the values of n for which the recurrence is true, which in this case means that we sum from $n = 1$ to $N - 1$. This yields

$$a \sum_{n=1}^{N-1} y_{n+1} x^n + b \sum_{n=1}^{N-1} y_n x^n + c \sum_{n=1}^{N-1} y_{n-1} x^n = \sum_{n=1}^{N-1} d_n x^n.$$

If we now express this equation in terms of our previously defined generating functions, it takes the form

$$\frac{a}{x}\{Y(x) - y_1 x\} + bY(x) + cx\{Y(x) - y_{N-1}x^{N-1}\} = D(x). \quad (1.4.2)$$

Next, with only a nagging doubt because y_1 and y_{N-1} are unknown, we press on with the recipe, whose next step asks us to solve (1.4.2) for the unknown generating function $Y(x)$. Now that isn't too hard, and we find at once that

$$\{a + bx + cx^2\}Y(x) = x\{D(x) + ay_1 + cy_{N-1}x^N\}. \qquad (1.4.3)$$

The unknown generating function $Y(x)$ is now known except for the two still-unknown constants y_1 and y_{N-1}, but (1.4.3) suggests a way to find them, too. There are two values of x, call them r_+ and r_-, at which the quadratic polynomial on the left side of (1.4.3) vanishes. Let us suppose that $r_+ \neq r_-$, for the moment. If we let $x = r_+$ in (1.4.3), we obtain one equation in the two unknowns y_1, y_{N-1}, and if we let $x = r_-$, we get another. The two equations are

$$\begin{aligned} ay_1 + (cr_+^N)y_{N-1} &= -D(r_+) \\ ay_1 + (cr_-^N)y_{N-1} &= -D(r_-). \end{aligned} \qquad (1.4.4)$$

Once these have been solved for y_1 and y_{N-1}, equation (1.4.3) then gives $Y(x)$ quite explicitly and completely. We leave the exceptional case where $r_+ = r_-$ to the reader.

1.5 Two independent variables

Until now we have studied sequences $\{x_n\}$ in *one* independent variable, n. In this section we begin the discussion of sequences in two independent variables, such as $\{x_{m,n}\}_{m,n \geq 0}$. In such cases it is often better to use the usual function notation $f(m, n)$ $(m, n \geq 0)$, where it is to be understood that the arguments are integer variables.

To illustrate the ideas in a simple context, we'll start with the binomial coefficients $\binom{n}{m}$. Combinatorially speaking, $\binom{n}{m}$ is the number of ways of choosing a subset of m letters from a set of n letters. More briefly, it is the number of m-subsets of $[n]$.*

The story in this section will be a little more extended than in earlier sections because, instead of regarding the recurrence relation as *given*, we are going to derive it by talking our way through it. This will be quite important because, in a great number of combinatorial situations, desired counting formulas will follow from using this same method.

So let's pretend that we don't know any formulas for the binomial coefficients, and we want to discover what the function $f(n, m)$ is, naturally by using the method of generating functions.

As we have defined it, $f(n, m)$ makes sense for all pairs of nonnegative integers. Note, for example, that $f(0, 0) = 1$, since the empty set has just

---

* We use the symbol $[n]$ to mean the set $\{1, 2, \ldots, n\}$.

one subset of 0 elements, namely itself! Note also that our definition implies that $f(n, m) = 0$ if $m > n$.

We extend the definition of f by requiring that $f(n, m) = 0$ if $n < 0$ or $m < 0$ (or both). Now f is defined for all pairs of integers, i.e., it is defined at every lattice point in the plane. The next job is to find a recurrence relation that f satisfies.

Fix a pair of nonnegative integers n, m, where $m < n$. Contemplate the collection of all of the m-subsets of $[n]$. There are $f(n, m)$ of them. Divide the collection into two piles, consisting of those subsets that *do* contain the highest letter n, and those that don't. How many subsets are there in each pile?

Well, if an m-subset does contain n, then if we delete n from it, it becomes an $(m - 1)$-subset of $[n - 1]$, and, conversely, if we choose some $(m - 1)$-subset of $[n - 1]$ and adjoin the singleton $\{n\}$ to it, we obtain an m-subset of $[n]$ that does contain n. *Hence the first of the two piles has exactly $f(n - 1, m - 1)$ subsets.*

On the other hand, if an m subset does not contain n, then it must be one of the $f(n - 1, m)$ m-subsets of $[n - 1]$. *Hence the second of the two piles has exactly $f(n - 1, m)$ subsets in it.*

Since every one of the $f(n, m)$ m-subsets of $[n]$ is in one or the other, but not both, of these piles, it must be true that

$$f(n, m) = f(n - 1, m) + f(n - 1, m - 1) \qquad (n, m = ?????????). \quad (1.5.1)$$

We are now well on the way to finding the recurrence relation that the 'unknown' function f satisfies. What remains before that task can be called complete is to replace the '?????????' in (1.5.1) by an exact description of the set of values of m and n for which the recurrence is true, because we will need to know that set quite precisely when it comes time to find generating functions.

OK, what *is* the set of m, n for which (1.5.1) is true? The argument by which we derived the formula, involving splitting up the subsets into two piles etc., proved the formula for certain pairs n, m. It surely did so for all pairs (n, m) for which $1 \le m \le n$, for then no negative numbers occur in either argument of f anywhere in (1.5.1). What if $m > n$? Then the left side of (1.5.1) is 0 and so are both terms on the right. Hence the formula is true for all $m, n \ge 1$. If m is negative, then all three terms shown are 0, which proves (1.5.1) for that case, and if $n < 0$ the same is true.

Now (1.5.1) has been proved unless either m or n is 0. Suppose $m = 0$ and $n \ne 0$. If n is positive, then the left side is 1 and so is the first term on the right, but the second term is 0, so (1.5.1) is true again. Similarly, if $n = 0$ and $m \ne 0$, both sides vanish.

This proves (1.5.1) for every lattice point in the plane with the possible exception of the origin $(n, m) = (0, 0)$. What happens there? Well, if $n = 0$

and $m = 0$, the left side of (1.5.1) is 1 and both terms on the right are 0, *so the statement is false in that case.**

To sum up what we know so far:

$$f(n,m) = f(n-1,m) + f(n-1,m-1) \qquad ((n,m) \neq (0,0); \; f(0,0) = 1).$$
$$(1.5.2)$$

That finishes the combinatorial part of the problem. What is left is the solution by generating functions. That process begins by defining the unknown generating function in terms of the members of the unknown sequence, as we have seen in the previous examples. In this case we have an unknown function of *two* variables, so at least three choices of generating functions might come to mind.

First, we might define a single generating function that has all of the unknown numbers in it, like

$$F(x,y) = \sum_{n,m} f(n,m)x^n y^m. \qquad (1.5.3)$$

Next, we might leave n as a free variable, and define a family of generating functions

$$G_n(y) = \sum_m f(n,m)y^m \qquad (\text{integer } n). \qquad (1.5.4)$$

Finally, we might leave m as a free variable and define the family of generating functions

$$H_m(x) = \sum_n f(n,m)x^n \qquad (\text{integer } m). \qquad (1.5.5)$$

All three of these are not only possible, they all lead to entertaining exercises, as well as to the right answers, so let's do all of them. It should be pointed out, though, that in most of our later work, one choice of generating function may give acceptable answers and others may not work at all.

We need to comment on the range of summation in the sums that appear in (1.5.3)-(1.5.5). We took some care to define $f(n,m)$ for all integers n, m, positive, negative, and zero. Therefore we will adopt the following conventions:

Convention 1. *When the range of a variable that is being summed over is not specified, it is to be understood that the range of summation is from $-\infty$ to $+\infty$.*

Convention 2. *When the range of a free variable in an equation is not specified, it is to be understood that the equation holds for all integer values of that variable.*

---

* The life of a working generatingfunctionologist is full of arguments like this one, so be prepared. We won't always work them out here in such detail, however.

These conventions will save us an enormous amount of work in the sequel, mainly in that we won't have to worry about changing the limits of summation when we change the variable of summation by a constant shift.

Let's work first with the choice (1.5.3) of the generating function. To do that we multiply (1.5.2) by $x^n y^m$, sum over all $(n,m) \neq (0,0)$, and obtain

$$F(x,y) - 1 = \sum_{n,m} f(n-1,m)x^n y^m + \sum_{n,m} f(n-1,m-1)x^n y^m.$$

In the first sum, take one factor of x outside the summation sign, then change the name of the dummy summation variable n, calling it $r+1$ instead. Similarly, in the second sum, take one factor of xy outside the summation sign, then change the name of both dummy summation variables n, m, calling them $r+1, s+1$ instead. The result is that

$$F(x,y) - 1 = x \sum_{r,m} f(r,m)x^r y^m + xy \sum_{r,s} f(r,s)x^r y^s \qquad (1.5.6)$$
$$= xF(x,y) + xyF(x,y).$$

We pause to exult in the fact that because of our conventions we didn't have to worry about changing limits of summation. When one takes a new dummy variable, like $n = r+1$, or some such, usually the upper and lower limits of the summation have to be changed up or down by one unit. Since we have arranged things so the limits are $\pm\infty$, that problem doesn't occur.

It is trivial to solve (1.5.6) for the unknown generating function

$$F(x,y) = \frac{1}{1-x-xy},$$

and therefore

$$f(n,m) = [x^n y^m]\frac{1}{1-x-xy}. \qquad (1.5.7)$$

Suppose instead that we choose to work with the generating functions (1.5.4). Then we would take a fixed $n \neq 0$, multiply (1.5.2) by y^m, and sum over all m. The result would be

$$G_n(y) = G_{n-1}(y) + yG_{n-1}(y) \qquad (n \neq 0),$$

from which each $G_n(y)$ is $(1+y)$ times the preceding $G_{n-1}(y)$. Since $G_0(y) = 1$ (why?), it follows that $G_n(y) = (1+y)^n$ $(n \geq 0)$, and therefore that

$$f(n,m) = [y^m](1+y)^n \qquad (n \geq 0). \qquad (1.5.8)$$

Similarly, if we work with the functions (1.5.5), we would fix an $m \neq 0$, multiply (1.5.2) by x^n and sum over all n, to get

$$H_m(x) = xH_m(x) + xH_{m-1}(x) \qquad (m \neq 0),$$

which yields $H_m(x) = (x/(1-x))H_{m-1}(x)$ for $m \geq 1$. But what is H_0? It is

$$H_0(x) = \sum_n f(n,0)x^n = \sum_{n \geq 0} x^n = \frac{1}{1-x}.$$

It follows that $H_m(x) = x^m/(1-x)^{m+1}$ for all $m \geq 0$, and therefore that

$$f(n,m) = [x^n]\left(\frac{x^m}{(1-x)^{m+1}}\right) \quad (m \geq 0)$$

$$= [x^{n-m}]\left(\frac{1}{1-x}\right)^{m+1}. \quad (1.5.9)$$

In (1.5.7)-(1.5.9) we have three generating functions for the binomial coefficients. How could we find formulas for those coefficients? Essentially, the question comes down to: how do we find the series expansion of a given function? Elementary calculus texts will assure you that the way to do that is to use Taylor's formula. That is, if $f(x)$ is given, and we want to write $f(x) = \sum_{n \geq 0} a_n x^n$, then to find the $\{a_n\}$ the thing to do is to write

$$a_n = \frac{1}{n!}\left(\frac{d}{dx}\right)^n f(x)\bigg|_{x=0} \quad (n \geq 0).$$

The problem with that is that the successive derivatives may become harder and harder to handle. For instance, to work out the Fibonacci numbers by repeated differentiation of the generating function $1/(1-x-x^2)$ is virtually out of the question, but partial fractions win the game quickly.

In the present case, Taylor's formula works just fine on one of the three forms in which we found the answer. In (1.5.9), it isn't hard at all to find the repeated derivatives of $(1-x)^{-m-1}$. In fact

$$\left(\frac{d}{dx}\right)^n\left\{\frac{1}{(1-x)^{m+1}}\right\} = \frac{(m+1)(m+2)\cdots(m+n)}{(1-x)^{m+n+1}}$$

$$= \frac{(m+n)!}{m!(1-x)^{m+n+1}} \quad (n \geq 0).$$

This means that

$$[x^n]\left(\frac{1}{(1-x)^{m+1}}\right) = \frac{(m+n)!}{m!n!} \quad (m,n \geq 0) \quad (1.5.10)$$

and so from (1.5.9),

$$f(n,m) = [x^{n-m}]\left(\frac{1}{1-x}\right)^{m+1}$$

$$= \frac{1}{(n-m)!}\frac{n!}{m!}$$

$$= \frac{n!}{m!(n-m)!} \quad (n \geq m \geq 0),$$

and we have found our explicit formula for the binomial coefficients.

If we had worked from (1.5.8), the calculation would have been about the same: Taylor's formula works, and in about the same way.

With (1.5.7), we would find the coefficients by using the geometric series first, to get

$$\frac{1}{1 - x - xy} = \sum_{r \geq 0} (x + xy)^r$$

$$= \sum_{r \geq 0} x^r (1 + y)^r,$$

from which

$$f(n, m) = [x^n y^m] \sum_{r \geq 0} x^r (1 + y)^r$$

$$= [y^m](1 + y)^n,$$

and we are back to (1.5.8) again.

Hence from all three of these generating functions we are able to extract the exact closed formula for the sequence of interest. We won't always be so lucky. Sometimes one approach will work and the others won't, while sometimes nothing seems to work.

1.6 Another 2-variable case

This example will have a stronger combinatorial flavor than the preceding ones. It concerns the partitions of a set. By a *partition of a set* S we will mean a collection of nonempty, pairwise disjoint sets whose union is S. Another name for a partition of S is an *equivalence relation* on S. The sets into which S is partitioned are called the *classes* of the partition.

For instance, we can partition $[5]^*$ in several ways. One of them is as $\{123\}\{4\}\{5\}$. In this partition there are three classes, one of which contains 1 and 2 and 3, another of which contains only 4, while the other contains only 5. No significance attaches to the order of the elements within the classes, nor to the order of the classes. All that matters is 'who is together and who is apart.'

Here is a list of *all* of the partitions of [4] into 2 classes:

$$\{12\}\{34\}; \ \{13\}\{24\}; \ \{14\}\{23\}; \ \{123\}\{4\}; \ \{124\}\{3\}; \ \{134\}\{2\}; \ \{1\}\{234\}.$$

$$(1.6.1)$$

There are exactly 7 partitions of [4] into 2 classes.

The problem that we will address in this example is to discover how many partitions of $[n]$ into k classes there are. Let $\left\{ {n \atop k} \right\}$ denote this number. It is called the Stirling number of the second kind. Our list above shows that $\left\{ {4 \atop 2} \right\} = 7$.

---

* Recall that $[n]$ is the set $\{1, 2, \ldots, n\}$

To find out more about these numbers we will follow the method of generating functions. First we will find a recurrence relation, then a few generating functions, then some exact formulas, etc.

We begin with a recurrence formula for $\left\{{n \atop k}\right\}$, and the derivation will be quite similar to the one used in the previous example.

Let positive integers n, k be given. Imagine that in front of you is the collection of all possible partitions of $[n]$ into k classes. There are exactly $\left\{{n \atop k}\right\}$ of them. As in the binomial coefficient example, we will carve up this collection into two piles; into the first pile go all of those partitions of $[n]$ into k classes in which the letter n lives in a class all by itself. Into the second pile go all other partitions, i.e., those in which the highest letter n lives in a class with other letters.

The question is, how many partitions are there in each of these two piles (expressed in terms of the Stirling numbers)?

Consider the first pile. There, every partition has n living alone. Imagine marching through that pile and erasing the class '(n)' that appears in every single partition in the pile. If that were done, then what would remain after the erasures is exactly the complete collection of all partitions of $[n-1]$ into $k-1$ classes. There are $\left\{{n-1 \atop k-1}\right\}$ of these, so there must have been $\left\{{n-1 \atop k-1}\right\}$ partitions in the first pile.

That was the easy one, but now consider the second pile. There the letter n always lives in a class with other letters. Therefore, if we march through that pile and erase the letter n wherever it appears, we won't affect the numbers of classes; we'll still be looking at partitions with k classes. After erasing the letter n from everything, our pile now contains partitions of $n-1$ letters into k classes. However, each one of these partitions appears not just once, but several times.

For example, in the list (1.6.1), the second pile contains the partitions

$$\{12\}\{34\};\ \{13\}\{24\};\ \{14\}\{23\};\ \{124\}\{3\};\ \{134\}\{2\};\ \{1\}\{234\}, \quad (1.6.2)$$

and after we delete '4' from every one of them we get the list

$$\{12\}\{3\};\ \{13\}\{2\};\ \{1\}\{23\};\ \{12\}\{3\};\ \{13\}\{2\};\ \{1\}\{23\}.$$

What we are looking at is the list of all partitions of $[3]$ into 2 classes, where each partition has been written down twice. Hence this list contains exactly $2\left\{{3 \atop 2}\right\}$ partitions.

In the general case, after erasing n from everything in the second pile, we will be looking at the list of all partitions of $[n-1]$ into k classes, where every such partition will have been written down k times. Hence that list will contain exactly $k\left\{{n-1 \atop k}\right\}$ partitions.

Therefore the second pile must have also contained $k\left\{{n-1 \atop k}\right\}$ partitions before the erasure of n.

The original list of $\left\{ {n \atop k} \right\}$ partitions was therefore split into two piles, the first of which contained $\left\{ {n-1 \atop k-1} \right\}$ partitions and the second of which contained $k\left\{ {n-1 \atop k} \right\}$ partitions.

It must therefore be true that

$$\left\{ {n \atop k} \right\} = \left\{ {n-1 \atop k-1} \right\} + k\left\{ {n-1 \atop k} \right\} \qquad ((n,k) = ????).$$

To determine the range of n and k, let's extend the definition of $\left\{ {n \atop k} \right\}$ to all pairs of integers. We put $\left\{ {n \atop k} \right\} = 0$ if $k > n$ or $n < 0$ or $k < 0$. Further, $\left\{ {n \atop 0} \right\} = 0$ if $n \neq 0$, and we will take $\left\{ {0 \atop 0} \right\} = 1$. With those conventions, the recurrence above is valid for all (n,k) other than $(0,0)$, and we have

$$\left\{ {n \atop k} \right\} = \left\{ {n-1 \atop k-1} \right\} + k\left\{ {n-1 \atop k} \right\} \qquad ((n,k) \neq (0,0); \left\{ {0 \atop 0} \right\} = 1). \qquad (1.6.3)$$

The stage is now set for finding the generating functions. Again there are three natural candidates for generating functions that might be findable, namely

$$A_n(y) = \sum_k \left\{ {n \atop k} \right\} y^k$$

$$B_k(x) = \sum_n \left\{ {n \atop k} \right\} x^n \qquad\qquad (1.6.4)$$

$$C(x,y) = \sum_{n,k} \left\{ {n \atop k} \right\} x^n y^k.$$

Before we plunge into the calculations, let's pause for a moment to develop some intuition about which of these choices is likely to succeed. To find $A_n(y)$ will involve multiplying (1.6.3) by y^k and summing over k. Do you see any problems with that? Well, there are some, and they arise from the factor of k in the second term on the right. Indeed, after multiplying by y^k and summing over k we will have to deal with something like $\sum_k k\left\{ {n \atop k} \right\} y^k$. This is certainly possible to think about, since it is related to the derivative of $A_n(y)$, but we do have a complication here.

If instead we choose to find $B_k(x)$, we multiply (1.6.3) by x^n and sum on n. Then the factor of k that seemed to be troublesome is not involved in the sum, and we can take that k outside of the sum as a multiplicative factor.

Comparing these, let's vote for the latter approach, and try to find the functions $B_k(x)$ $(k \geq 0)$. Hence, multiply (1.6.3) by x^n and sum over n, to get

$$B_k(x) = x B_{k-1}(x) + kx B_k(x) \qquad (k \geq 1; B_0(x) = 1).$$

This leads to

$$B_k(x) = \frac{x}{1-kx} B_{k-1}(x) \qquad (k \geq 1; B_0(x) = 1)$$

and finally to the evaluation

$$B_k(x) = \sum_n \left\{ {n \atop k} \right\} x^n = \frac{x^k}{(1-x)(1-2x)(1-3x)\cdots(1-kx)} \quad (k \geq 0).$$

$$(1.6.5)$$

The problem of finding an explicit formula for the Stirling numbers could therefore be solved if we could find the power series expansion of the function that appears in (1.6.5). That, in turn, calls for a dose of partial fractions, *not* of Taylor's formula!

The partial fraction expansion in question has the form

$$\frac{1}{(1-x)(1-2x)\cdots(1-kx)} = \sum_{j=1}^k \frac{\alpha_j}{(1-jx)}.$$

To find the α's, fix r, $1 \leq r \leq k$, multiply both sides by $1 - rx$, and let $x = 1/r$. The result is that

$$\alpha_r = \frac{1}{(1-1/r)(1-2/r)\cdots(1-(r-1)/r)(1-(r+1)/r)\cdots(1-k/r)}$$

$$= (-1)^{k-r} \frac{r^{k-1}}{(r-1)!(k-r)!} \quad (1 \leq r \leq k).$$

$$(1.6.6)$$

From (1.6.5) and (1.6.6) we obtain, for $n \geq k$,

$$\left\{ {n \atop k} \right\} = [x^n] \left\{ \frac{x^k}{(1-x)(1-2x)\cdots(1-kx)} \right\}$$

$$= [x^{n-k}] \left\{ \frac{1}{(1-x)(1-2x)\cdots(1-kx)} \right\}$$

$$= [x^{n-k}] \sum_{r=1}^k \frac{\alpha_r}{1-rx} \quad (k \geq 1)$$

mult by denom

$$= \sum_{r=1}^k \alpha_r [x^{n-k}] \frac{1}{1-rx}$$

$$(1.6.7)$$

$$= \sum_{r=1}^k \alpha_r r^{n-k}$$

$$= \sum_{r=1}^k (-1)^{k-r} \frac{r^{k-1}}{(r-1)!(k-r)!} r^{n-k}$$

$$= \sum_{r=1}^k (-1)^{k-r} \frac{r^n}{r!(k-r)!} \quad (n, k \geq 0),$$

which is just what we wanted: an explicit formula for $\left\{{n \atop k}\right\}$. Do check that this formula yields $\left\{{4 \atop 2}\right\} = 7$, which we knew already. At the same time, note that the formula says that $\left\{{n \atop 2}\right\} = 2^{n-1} - 1$ $(n > 0)$. Can you give an independent proof of that fact?

Next, we're going to try one of the other approaches to solving the recurrence for the Stirling numbers, namely that of studying the functions $A_n(y)$ in (1.6.4). This method is much harder to carry out to completion than the one we just used, but it turns out that these generating functions have other uses that are quite important from a theoretical point of view.

Therefore, let's fix $n > 0$, multiply (1.6.3) by y^k and sum over k. The result is

$$
\begin{aligned}
A_n(y) &= \sum_k \left\{{n-1 \atop k-1}\right\} y^k + \sum_k k \left\{{n-1 \atop k}\right\} y^k \\
&= y A_{n-1}(y) + (y \frac{d}{dy}) A_{n-1}(y) \\
&= \{y(1 + D_y)\} A_{n-1}(y) \qquad (n > 0; A_0(y) = 1).
\end{aligned}
\tag{1.6.8}
$$

The novel feature is the appearance of the differentiation operator d/dy that was necessitated by the factor k in the recurrence relation.

Hence each function A_n is obtained from its predecessor by applying the operator $y(1 + D_y)$. Beginning with $A_0 = 1$, we obtain successively $y, y + y^2, y + 3y^2 + y^3, \ldots$, but as far as an explicit formula is concerned, we find only that

$$
A_n(y) = \{y + y D_y\}^n 1 \qquad (n \geq 0)
\tag{1.6.9}
$$

by this approach.

There are, however, one or two things that can be seen more clearly from these generating functions than from the $B_n(x)$'s. One of these will be discussed in section 4.5, and concerns the *shape* of the sequence $\left\{{n \atop k}\right\}$ for fixed n, as k runs from 1 to n. It turns out that the sequence increases for a while and then it decreases. That is, it has just one maximum. Many combinatorial sequences are *unimodal*, like this one, but in some cases it can be very hard to prove such things. In this case, thanks to the formula (1.6.9), we will see that it's not hard at all.

For an application of (1.6.7), recall that the Stirling number $\left\{{n \atop k}\right\}$ is the number of ways of partitioning a set of n elements into k classes. Suppose we don't particularly care how many classes there are, but we want to know the number of ways to partition a set of n elements. Let these numbers be $\{b(n)\}_0^\infty$. They are called the Bell numbers. It is conventional to take $b(0) = 1$. The sequence of Bell numbers begins 1, 1, 2, 5, 15, 52, $\ldots$

Can we find an explicit formula for the Bell numbers? Nothing to it. In (1.6.7) we have an explicit formula for $\left\{{n \atop k}\right\}$. If we sum that formula from $k = 1$ to n we will have an explicit formula for $b(n)$. *However*, there's one more thing that it is quite profitable to notice. The formula (1.6.7) is valid for *all* positive integer values of n and k. In particular, it is valid if $k > n$. But $\left\{{n \atop k}\right\} = 0$ if $k > n$. This means that the formula (1.6.7) doesn't have to be *told* that $\left\{{13 \atop 19}\right\} = 0$; it *knows* it; i.e., if we blissfully insert $n = 13$, $k = 19$ into the monster sum and work it all out, we will get 0.

Hence, to calculate the Bell numbers, we can sum the last member of (1.6.7) from $k = 1$ to M, where M is any number you please that is $\geq n$. Let's do it. The result is that

$$b(n) = \sum_{k=1}^{M} \sum_{r=1}^{k} (-1)^{k-r} \frac{r^{n-1}}{(r-1)!(k-r)!}$$

$$= \sum_{r=1}^{M} \frac{r^{n-1}}{(r-1)!} \sum_{k=r}^{M} \frac{(-1)^{k-r}}{(k-r)!}$$

$$= \sum_{r=1}^{M} \frac{r^{n-1}}{(r-1)!} \left\{ \sum_{s=0}^{M-r} \frac{(-1)^s}{s!} \right\}.$$

But now the number M is arbitrary, except that $M \geq n$. Since the partial sum of the exponential series in the curly braces above is so inviting, let's keep n and r fixed, and let $M \to \infty$. This gives the following remarkable formula for the Bell numbers (check it yourself for $n = 1$):

$$b(n) = \frac{1}{e} \sum_{r \geq 0} \frac{r^n}{r!} \qquad (n \geq 0). \tag{1.6.10}$$

This formula for the Bell numbers, although it has a certain charm, doesn't lend itself to computation. From it, however, we can derive a generating function for the Bell numbers that is unexpectedly simple and elegant. We will look for the generating function in the form

$$B(x) = \sum_{n \geq 0} \frac{b(n)}{n!} x^n. \tag{1.6.11}$$

This is the first time we have found it necessary to introduce an extra factor of $1/n!$ into the coefficients of a generating function. That kind of thing happens frequently, however, and we will discuss in chapter 2 how to recognize when extra factors like these will be useful. A generating function of the form (1.6.11), with the $1/n!$'s thrown into the coefficients, is called an *exponential generating function*. We would say, for instance, that '$B(x)$ is the exponential generating function of the Bell numbers.'

When we wish to distinguish the various kinds of generating functions, we may use the phrase *the ordinary power series generating function of the sequence $\{a_n\}$ is $\sum_n a_n x^n$* or *the exponential generating function of the sequence $\{a_n\}$ is $\sum_n a_n x^n / n!$*.

To find $B(x)$ explicitly, take the formula (1.6.10), which is valid for $n \geq 1$, multiply it by $x^n/n!$ (don't forget the $n!$), and sum over all $n \geq 1$. This gives

$$B(x) - 1 = (\frac{1}{e}) \sum_{n \geq 1} \frac{x^n}{n!} \sum_{r \geq 1} \frac{r^{n-1}}{(r-1)!}$$

$$= (\frac{1}{e}) \sum_{r \geq 1} \frac{1}{r!} \sum_{n \geq 1} \frac{(rx)^n}{n!}$$

$$= (\frac{1}{e}) \sum_{r \geq 1} \frac{1}{r!}(e^{rx} - 1)$$

$$= (\frac{1}{e})\{e^{e^x} - e\}$$

$$= e^{e^x - 1} - 1.$$

We have therefore shown

Theorem 1.6.1. *The exponential generating function of the Bell numbers is $e^{e^x - 1}$, i.e., the coefficient of $x^n/n!$ in the power series expansion of $e^{e^x - 1}$ is the number of partitions of a set of n elements.*

This result is surely an outstanding example of the power of the generating function approach. The Bell numbers themselves are complicated, but the generating function is simple and easy to remember.

The next novel element of this story is the fact that we can go *from* generating functions *to* recurrence formulas, although in all our examples to date the motion has been in the other direction. We propose now to derive from Theorem 1.6.1 a recurrence formula for the Bell numbers, one that will make it easy to compute as many of them as we might wish to look at.

First, the theorem tells us that

$$\sum_{n \geq 0} \frac{b(n)}{n!} x^n = e^{e^x - 1}. \tag{1.6.12}$$

We are going to carry out a very standard operation on this equation, but the first time this operation appears it seems to be anything but standard.

The $x(d/dx) \log$ operation

(1) Take the logarithm of both sides of the equation.
(2) Differentiate both sides and multiply through by x.
(3) Clear the equation of fractions.
(4) For each n, find the coefficients of x^n on both sides of the equation and equate them.

Although the best motivation for the above program is the fact that it works, let's pause for a moment before doing it, to see why it is likely to work. The point of taking logarithms is to simplify the function e^{e^x-1}, whose power series coefficients are quite mysterious before taking logarithms, and are quite obvious after doing so. The price for that simplification is that on the left side we have the log of a sum, which is an awesome thing to have. The next step, the differentiation, changes the log of the sum into a ratio of two sums, which is much nicer. The reason for multiplying through by x is that the differentiation dropped the power of x by 1 and it's handy to restore it. After clearing of fractions we will simply be looking at two sums that are equal to each other, and the work will be over.

In this case, after step 1 is applied to (1.6.12), we have

$$\log\left\{\sum_{n\geq 0}\frac{b(n)}{n!}x^n\right\} = e^x - 1.$$

Step 2 gives

$$\frac{\sum_n\frac{nb(n)x^n}{n!}}{\sum_n\frac{b(n)x^n}{n!}} = xe^x.$$

To clear of fractions, multiply both sides by the denominator on the left, obtaining

$$\sum_n\frac{nb(n)x^n}{n!} = (xe^x)\sum_n\frac{b(n)x^n}{n!}.$$

Finally, we have to identify the coefficients of x^n on both sides of this equation. On the left it's easy. On the right we have to multiply two power series together first, and then identify the coefficient. Since in chapter 2 we will work out a general and quite easy-to-use rule for doing things like this, let's postpone this calculation until then, and merely quote the result here. It is that the Bell numbers satisfy the recurrence

$$b(n) = \sum_k \binom{n-1}{k}b(k) \qquad (n\geq 1; b(0)=1). \tag{1.6.13}$$

We have now seen several examples of how generating functions can be used to find recurrence relations. It often happens that the method of generating functions finds a recurrence, and only later are we able to give a direct, combinatorial interpretation of the recurrence. In some cases, recurrences are known that look like they ought to have simple combinatorial explanations, but none have yet been found.

Exercises

1. Find the ordinary power series generating functions of each of the following sequences, in simple, closed form. In each case the sequence is defined for all $n \geq 0$.

    (a) $a_n = n$

    (b) $a_n = \alpha n + \beta$

    (c) $a_n = n^2$

    (d) $a_n = \alpha n^2 + \beta n + \gamma$

    (e) $a_n = P(n)$, where P is a given polynomial, of degree m.

    (f) $a_n = 3^n$

    (g) $a_n = 5 \cdot 7^n - 3 \cdot 4^n$

2. For each of the sequences given in part 1, find the exponential generating function of the sequence in simple, closed form.

3. If $f(x)$ is the ordinary power series generating function of the sequence $\{a_n\}_{n \geq 0}$, then express simply, in terms of $f(x)$, the ordinary power series generating functions of the following sequences. In each case the range of n is $0, 1, 2, \ldots$

    (a) $\{a_n + c\}$

    (b) $\{\alpha a_n + c\}$

    (c) $\{n a_n\}$

    (d) $\{P(n)a_n\}$, where P is a given polynomial.

    (e) $0, a_1, a_2, a_3, \ldots$

    (f) $0, 0, 1, a_3, a_4, a_5, \ldots$

    (g) $a_0, 0, a_2, 0, a_4, 0, a_6, 0, a_8, 0, \ldots$

    (h) $a_1, a_2, a_3, \ldots$

    (i) $\{a_{n+h}\}$ (h a given constant)

    (j) $\{a_{n+2} + 3a_{n+1} + a_n\}$

    (k) $\{a_{n+2} - a_{n+1} - a_n\}$

4. Let $f(x)$ be the *exponential* generating function of a sequence $\{a_n\}$. For each of the sequences in exercise 3, find the exponential generating function simply, in terms of $f(x)$.

5. Find

(a) $[x^n]e^{2x}$

(b) $[x^n/n!]e^{\alpha x}$

(c) $[x^n/n!]\sin x$

(d) $[x^n]\{1/((1-ax)(1-bx))\}$ $(a \neq b)$

(e) $[x^n](1+x^2)^m$

6. In each part, a sequence $\{a_n\}_{n \geq 0}$ satisfies the given recurrence relation. Find the ordinary power series generating function of the sequence.

(a) $a_{n+1} = 3a_n + 2$ $(n \geq 0; a_0 = 0)$

(b) $a_{n+1} = \alpha a_n + \beta$ $(n \geq 0; a_0 = 0)$

(c) $a_{n+2} = 2a_{n+1} - a_n$ $(n \geq 0; a_0 = 0; a_1 = 1)$

(d) $a_{n+1} = a_n/3 + 1$ $(n \geq 0; a_0 = 0)$

7. Give a direct combinatorial proof of the recurrence (1.6.13), as follows: given n; consider the collection of all partitions of the set $[n]$. There are $b(n)$ of them. Sort out this collection into piles numbered $k = 0, 1, \ldots, n-1$, where the kth pile consists of all partitions of $[n]$ in which the class that contains the letter 'n' contains exactly k other letters. Count the partitions in the kth pile, and you'll be all finished.

8. In each part of problem 6, find the exponential generating function of the sequence (you may have to solve a differential equation to do so!).

9. A function f is defined for all $n \geq 1$ by the relations (a) $f(1) = 1$ and (b) $f(2n) = f(n)$ and (c) $f(2n+1) = f(n) + f(n+1)$. Let

$$F(x) = \sum_{n \geq 1} f(n)x^{n-1}$$

be the generating function of the sequence. Show that

$$F(x) = (1 + x + x^2)F(x^2),$$

and therefore that

$$F(x) = \prod_{j \geq 0}^{\infty} \left\{1 + x^{2^j} + x^{2^{j+1}}\right\}.$$

10. Let X be a random variable that takes the values $0, 1, 2, \ldots$ with respective probabilities $p_0, p_1, p_2, \ldots$, where the p's are given nonnegative real numbers whose sum is 1. Let $P(x)$ be the opsgf of $\{p_n\}$.

(a) Express the mean μ and standard deviation σ of X directly in terms of $P(x)$.

(b) Two values of X are sampled independently. What is the probability $p_n^{(2)}$ that their sum is n? Express the opsgf $P_2(x)$ of $\{p_n^{(2)}\}$ in terms of $P(x)$.

(c) k values of X are sampled independently. Let $p_n^{(k)}$ be the probability that their sum is equal to n. Express the opsgf $P_k(x)$ of $\{p_n^{(k)}\}_{n\geq 0}$ in terms of $P(x)$.

(d) Use the results of parts (a) and (c) to find the mean and standard deviation of the sum of k independently chosen values of X, in terms of μ and σ.

(e) Let $A(x)$ be a power series with $A(0) = 1$, and let $B(x) = A(x)^k$. It is desired to compute the coefficients of $B(x)$, without raising $A(x)$ to any powers at all. Use the '$xD\log$' method to derive a recurrence formula that is satisfied by the coefficients of $B(x)$.

(f) A loaded die has probabilities .1, .2, .1, .2, .2, .2 of turning up with, respectively, 1, 2, 3, 4, 5, or 6 spots showing. The die is then thrown 100 times, and we want to calculate the probability p^* that the total number of spots on all 100 throws is ≤ 300. Identify p^* as the coefficient of x^{300} in the power series expansion of a certain function. Say exactly what the function is (you are *not* being asked to calculate p^*). Use the result of part (e) to say exactly how you would calculate p^* if you had to.

(g) A random variable X assumes each of the values $1, 2, \ldots, m$ with probability $1/m$. Let S_n be the result of sampling n values of X independently and summing them. Show that for $n = 1, 2, \ldots$,

$$\text{Prob}\{S_n \leq j\} = \frac{1}{m^n}\sum_r(-1)^r\binom{n}{r}\binom{j-mr}{n}.$$

11. Let $f(n)$ be the number of subsets of $[n]$ that contain no two consecutive elements, for integer n. Find the recurrence that is satisfied by these numbers, and then 'find' the numbers themselves.

12. For given integers n, k, let $f(n, k)$ be the number of k-subsets of $[n]$ that contain no two consecutive elements. Find the recurrence that is satisfied by these numbers, find a suitable generating function and find the numbers themselves. Show the numerical values of $f(n, k)$ in a Pascal triangle arrangement, for $n \leq 6$.

13. By comparing the results of the above two problems, deduce an identity. Draw a picture of the elements of Pascal's triangle that are involved in this identity.

14. Let the integers $1, 2, \ldots, n$ be arranged consecutively around a circle, and let $g(n)$ be the number of ways of choosing a subset of these, no two

consecutive on the circle. That is, g differs from the f of problem 11 in that n and 1 are now regarded as consecutive. Find $g(n)$.

15. As in the previous problem, find, analogously to problem 12 above, the number $g(n,k)$ of ways of choosing k elements from n arranged on a circle, such that no two chosen elements are adjacent on the circle.

16. Find the coefficient of x^n in the power series for

$$f(x) = \frac{1}{(1-x^2)^2},$$

first by the method of partial fractions, and second, give a much simpler derivation by being sneaky.

17. An *inversion* of a permutation σ of $[n]$ is a pair of letters i, j such that $i < j$ and $\sigma(i) > \sigma(j)$. In the 2-line form of writing the permutation, an inversion shows up as a pair that is 'in the wrong order' in the second line. The permutation

$$\sigma = \begin{pmatrix} 1 & 2 & 3 & 4 & 5 & 6 & 7 & 8 & 9 \\ 4 & 9 & 2 & 5 & 8 & 1 & 6 & 7 & 3 \end{pmatrix}$$

of $[9]$ has 19 inversions, namely the pairs (4,2), (4,1), ..., (7,3). Let $b(n,k)$ be the number of permutations of n letters that have exactly k inversions. Find a 'simple' formula for the generating function $B_n(x) = \sum_k b(n,k)x^k$. Make a table of values of $b(n,k)$ for $n \le 5$.

18.

(a) Given n, k. For how many of the permutations of n letters is it true that their first k values decrease?

(b) What is the average length of the decreasing sequence with which the values of a random n-permutation begin?

(c) If $f(n,k)$ is the number of permutations that have exactly k ascending runs, find the Pascal-triangle-type recurrence satisfied by $f(n,k)$. They are called the Euler numbers. As an example, the permutation

$$\begin{pmatrix} 1 & 2 & 3 & 4 & 5 & 6 & 7 & 8 & 9 \\ 4 & 1 & 6 & 9 & 2 & 5 & 8 & 3 & 7 \end{pmatrix}$$

has 4 such runs, namely 4, 1 6 9, 2 5 8, and 3 7.

19. Consider the 256 possible sums of the form

$$\epsilon_1 + \epsilon_2 + 2\epsilon_3 + 5\epsilon_4 + 10\epsilon_5 + 10\epsilon_6 + 20\epsilon_7 + 50\epsilon_8 \tag{1}$$

where each ϵ is 0 or 1.

(a) For each integer n, let C_n be the number of different sums that represent n. Write the generating polynomial

$$C_0 + C_1 x + C_2 x^2 + C_3 x^3 + \cdots + C_{99} x^{99}$$

as a product.

(b) Next, consider all of the possible sums that are formed as in (1), where now the ϵ's can have any of the three values $-1, 0, 1$. For each integer n, let D_n be the number of different sums that represent n. Show that some integer n is representable in at least 33 different ways. Then write the generating function

$$\sum_{n=-99}^{99} D_n x^n$$

as a product.

(c) Generalize the results of parts (a) and (b) of this problem by replacing the particular set of weights by a general set. Factor the polynomial that occurs.

(d) In the general case of part (d) of this problem, state precisely what all of the zeros of the generating polynomial are, and state precisely what the multiplicity of each of the zeros is, in terms of the set of weights.

20. Let $f(n, m, k)$ be the number of strings of n 0's and 1's that contain exactly m 1's, no k of which are consecutive.

(a) Find a recurrence formula for f. It should have $f(n, m, k)$ on the left side, and *exactly three terms on the right*.

(b) Find, in simple closed form, the generating functions

$$F_k(x, y) = \sum_{n,m \geq 0} f(n, m, k) x^n y^m \qquad (k = 1, 2, \ldots).$$

(c) Find an explicit formula for $f(n, m, k)$ from the generating function (this should involve only a single summation, of an expression that involves a few factorials).

21.

(a) We want to find a formula for the nth derivative of the function e^{e^x}. Differentiate it a few times, study the pattern, and conjecture the form of the answer for general n, including some constants to be determined. Then find a recurrence formula for the constants in question, and identify them as some 'famous' numbers that we have studied.

(b) Next let $f(x_1, \ldots, x_n)$ be some function of n variables. Find a formula for the mixed partial derivative

$$\frac{\partial^n}{\partial x_1 \partial x_2 \cdots \partial x_n} e^f$$

that expresses it in terms of various partial derivatives of f itself.

Chapter 2
Series

This chapter is devoted to a study of the different kinds of series that are widely used as generating functions.

2.1 Formal power series

To discuss the *formal* theory of power series, as opposed to their *analytic* theory, is to discuss these series as purely algebraic objects, in their roles as clotheslines, without using any of the function-theoretic properties of the function that may be represented by the series or, indeed, without knowing whether such a function exists.

We study formal series because it often happens in the theory of generating functions that we are trying to solve a recurrence relation, so we introduce a generating function, and then we go through the various manipulations that follow, but with a guilty conscience because we aren't sure whether the various series that we're working with will converge. Also, we might find ourselves working with the derivatives of a generating function, still without having any idea if the series converges to a function at all.

The point of this section is that there's no need for the guilt, because the various manipulations can be carried out in the ring of formal power series, where questions of convergence are nonexistent. We may execute the whole method and end up with the generating series, and only then discover whether it converges and thereby represents a real honest function or not. If not, we may still get lots of information from the formal series, but maybe we won't be able to get *analytic* information, such as asymptotic formulas for the sizes of the coefficients. Exact formulas for the sequences in question, however, might very well still result, even though the method rests, in those cases, on a purely algebraic, formal foundation.

The series

$$f = 1 + x + 2x^2 + 6x^3 + 24x^4 + 120x^5 + \cdots + n!x^n + \cdots, \qquad (2.1.1)$$

for instance, has a perfectly fine existence as a formal power series, despite the fact that it converges for no value of x other than $x = 0$, and therefore offers no possibilities for investigation by analytic methods. Not only that, but this series plays an important role in some natural counting problems.

A *formal power series* is an expression of the form

$$a_0 + a_1 x + a_2 x^2 + \cdots$$

where the sequence $\{h_n\}_0^\infty$ is called the *sequence of coefficients*. To say that two series are *equal* is to say that their coefficient sequences are the same.

We can do certain kinds of operations with formal power series. We can *add* or *subtract* them, for example. This is done according to the rules

$$\sum_n a_n x^n \pm \sum_n b_n x^n = \sum_n (a_n \pm b_n) x^n.$$

Power series can be multiplied by the usual Cauchy product rule,

$$\sum_n a_n x^n \sum_n b_n x^n = \sum_n c_n x^n \qquad (c_n = \sum_k a_k b_{n-k}). \qquad (2.1.2)$$

It is certainly this product rule that accounts for the wide applicability of series methods in combinatorial problems. This is because frequently we can construct all a_n of the objects of type n in some family by choosing an object of type k and an object of type $n - k$ and stitching them together to make the object of type n. The number of ways of doing that will be $a_k a_{n-k}$, and if we sum on k we find that the Cauchy product of two formal series is directly relevant to the problem that we are studying.

If we follow the multiplication rule we obtain, for instance,

$$(1 - x)(1 + x + x^2 + x^3 + \cdots) = 1.$$

Thus we can say that the series $(1 - x)$ has a reciprocal, and that reciprocal is $1 + x + x^2 + \cdots$ (and the other way around, too).

Proposition. *A formal power series* $f = \sum_{n \geq 0} a_n x^n$ *has a reciprocal if and only if* $a_0 \neq 0$. *In that case the reciprocal is unique.*

Proof. Let f have a reciprocal, namely $1/f = \sum_{n \geq 0} b_n x^n$. Then $f \cdot (1/f) = 1$ and according to (2.1.2), $c_0 = 1 = a_0 b_0$, so $a_0 \neq 0$. Further, in this case (2.1.2) tells us that for $n \geq 1$, $c_n = 0 = \sum_k a_k b_{n-k}$, from which we find

$$b_n = (-1/a_0) \sum_{k \geq 1} a_k b_{n-k} \qquad (n \geq 1). \qquad (2.1.3)$$

This determines $b_1, b_2, \ldots$ uniquely, as claimed.

Conversely, suppose $a_0 \neq 0$. Then we can determine $b_0, b_1, \ldots$ from (2.1.3), and the resulting series $\sum_n b_n x^n$ is the reciprocal of f. ∎

The collection of formal power series under the rules of arithmetic that we have just described forms a *ring*, in which the invertible elements are the series with nonvanishing constant term.

The above idea of a *reciprocal* of a formal power series is not to be confused with the subtler notion of the *inverse* of such a series. The inverse of a series f, if it exists, is a series g such that $f(g(x)) = g(f(x)) = x$. When can such an inverse exist? First we need to be able to *define* the symbol $f(g(x))$, then we can worry about whether or not it is equal to x.

If $f = \sum_n a_n x^n$, then $f(g(x))$ means

$$f(g(x)) = \sum_n a_n g(x)^n. \qquad (2.1.4)$$

If the series $g(x)$ has a nonzero constant term, g_0, then every term of the series (2.1.4) may contribute to the coefficient of each power of x. On the other hand, if $g_0 = 0$, then we will be able to compute the coefficient of, say, x^{57} in (2.1.4) from just the first 58 terms of the series shown. Indeed, notice that every single term

$$a_n g(x)^n = a_n (g_1 x + g_2 x^2 + \ldots)^n$$
$$= a_n x^n (g_1 + g_2 x + \ldots)^n$$

with $n > 57$ will contain only powers of x higher than the 57th, and therefore we won't need to look at those terms to find the coefficient of x^{57}.

Thus if $g_0 = 0$ then the computation of each one of the coefficients of the series $f(g(x))$ is a *finite* process, and therefore all of those coefficients are well defined, and so is the series. If $g_0 \neq 0$, though, the computation of each coefficient of $f(g(x))$ is an *infinite* process unless f is a polynomial, and therefore it will make sense only if the series 'converge.' In a formal, algebraic theory, however, ideas of convergence have no place. Thus *the composition $f(g(x))$ of two formal power series is defined if and only if $g_0 = 0$ or f is a polynomial.*

For instance, the series $e^{e^x - 1}$ is a well defined *formal* series, whereas the series e^{e^x} is not defined, at least from the general definition of composition of functions.

To return to the question of finding a series inverse of a given series f, we see that if such an inverse series g exists, then

$$f(g(x)) = g(f(x)) = x \qquad (2.1.5)$$

must both make sense and be true. We claim that if $f(0) = 0$ *the inverse series exists if and only if the coefficient of x is nonzero in the series f.*

Proposition. *Let the formal power series f, g satisfy (2.1.5) and $f(0) = 0$. Then $f = f_1 x + f_2 x^2 + \cdots$ ($f_1 \neq 0$), and $g = g_1 x + g_2 x^2 + \cdots$ ($g_1 \neq 0$).*

Proof. Suppose that $f = f_r x^r + \cdots$ and $g = g_s x^s + \cdots$, where $r, s \geq 0$ and $f_r g_s \neq 0$. Then $f(g(x)) = x = f_r g_s^r x^{rs} + \cdots$, whence $rs = 1$, and $r = s = 1$, as claimed. ∎

In the ring of formal power series there are other operations defined, which mirror the corresponding operations of function calculus, but which make no use of limiting operations.

The *derivative* of the formal power series $f = \sum_n a_n x^n$ is the series $f' = \sum_n n a_n x^{n-1}$. Differentiation follows the usual rules of calculus, such as the sum, product, and quotient rules. Many of these properties are even easier to prove for formal series than they are for the functions of calculus. For example:

Proposition. *If $f' = 0$ then $f = a_0$ is constant.*

Proof. Take another look at the '=' sign in the hypothesis $f' = 0$. It means that the formal power series f' is identical to the formal power series 0, and that means that each and every coefficient of the formal series f' is 0. But the coefficients of f' are $a_1, 2a_2, 3a_3, \ldots$, so each of these is 0, and therefore $a_j = 0$ for all $j \geq 1$, which is to say that f is constant. ∎

Next, try this one:

Proposition. *If $f' = f$ then $f = ce^x$.*

Proof. Since $f' = f$, the coefficient of x^n must be the same in f as in f', for all $n \geq 0$. Hence $(n + 1)a_{n+1} = a_n$ for all $n \geq 0$, whence $a_{n+1} = a_n/(n + 1)$ $(n \geq 0)$. By induction on n, $a_n = a_0/n!$ for all $n \geq 0$, and so $f = a_0 e^x$. ∎

2.2 The calculus of formal ordinary power series generating functions

Operations on formal series involve corresponding operations on their coefficients. If the series actually converge and represent functions, then operations on those functions correspond to certain operations on the power series coefficients of the expansions of those functions. In this section we will explore some of these relationships. They are of great importance in helping to spot which kind of generating function is appropriate for which kind of recurrence relation or other combinatorial situation.

Definition. *The symbol $f \overset{ops}{\longleftrightarrow} \{h_n\}_0^\infty$ means that the series f is the ordinary power series ('ops') generating function for the sequence $\{h_n\}_0^\infty$. That is, it means that $f = \sum_n a_n x^n$.*

Suppose $f \overset{ops}{\longleftrightarrow} \{h_n\}_0^\infty$. Then what generates $\{a_{n+1}\}_0^\infty$? To answer that we do a little calculation:

$$\sum_{n \geq 0} a_{n+1} x^n = \frac{1}{x} \sum_{m \geq 1} a_m x^m = \frac{(f(x) - f(0))}{x}.$$

Therefore

$$f \overset{ops}{\longleftrightarrow} \{h_n\}_0^\infty \Rightarrow ((f - a_0)/x) \overset{ops}{\longleftrightarrow} \{a_{n+1}\}_0^\infty. \tag{2.2.1}$$

Thus a shift of the subscript by 1 unit changes the series represented to the difference quotient $(f - a_0)/x$. If we shift by 2 units, of course, we just iterate the difference quotient operation, and find that

$$\{a_{n+2}\}_0^\infty \overset{ops}{\longleftrightarrow} \frac{((f - a_0)/x) - a_1}{x}$$

$$= \frac{f - a_0 - a_1 x}{x^2}.$$

Note how this point of view allows us to see 'at a glance' that the Fibonacci recurrence relation $F_{n+2} = F_{n+1} + F_n$ $(n \geq 0; F_0 = 0; F_1 = 1)$ translates directly into the ordinary power series generating function relation

$$\frac{f - x}{x^2} = \frac{f}{x} + f.$$

Indeed, the purpose of this section is to develop this facility for passing from *sequence* relations to *series* relations quickly and conveniently.

Rule 1. *If* $f \overset{ops}{\longleftrightarrow} \{h_n\}_0^\infty$, *then, for integer* $h > 0$,

$$\{a_{n+h}\}_0^\infty \overset{ops}{\longleftrightarrow} \frac{f - a_0 - \cdots - a_{h-1}x^{h-1}}{x^h}.$$

Next let's look into the effect of multiplying the sequence by powers of n. Again, suppose that $f \overset{ops}{\longleftrightarrow} \{h_n\}_0^\infty$. Then what generates the sequence $\{na_n\}_0^\infty$? The question means this: can we express the series $\sum_n na_n x^n$ in some simple way in terms of the series $f = \sum_n a_n x^n$? The answer is easy, because the former series is exactly xf'. Therefore, *to multiply the nth member of a sequence by n causes its ops generating function to be 'multiplied' by* $x(d/dx)$, which we will write as xD. In symbols:

$$f \overset{ops}{\longleftrightarrow} \{h_n\}_0^\infty \Rightarrow (xDf) \overset{ops}{\longleftrightarrow} \{na_n\}_0^\infty. \qquad (2.2.2)$$

As an example, consider the recurrence

$$(n+1)a_{n+1} = 3a_n + 1 \qquad (n \geq 0; a_0 = 1).$$

If f is the opsgf of the sequence $\{h_n\}_0^\infty$, then from Rule 1 and (2.2.2),

$$f' = 3f + \frac{1}{1-x},$$

which is a first order differential equation in the unknown generating function, and it can be solved by standard methods.

Next suppose $f \overset{ops}{\longleftrightarrow} \{h_n\}_0^\infty$. Then what generates the sequence

$$\{n^2 a_n\}_0^\infty?$$

Obviously we re-apply the multiply-by-n operator xD, so the answer is $(xD)^2 f$. In general,

$$(xD)^k f \overset{ops}{\longleftrightarrow} \{n^k a_n\}_{n\geq 0}.$$

OK, what generates $\{(3 - 7n^2)a_n\}_{n\geq 0}$? Again obviously, we do the same thing to xD that is done to n, i.e., $(3 - 7(xD)^2)f$ is the answer. The general prescription is:

Rule 2. *If* $f \overset{ops}{\longrightarrow} \{h_n\}_0^\infty$, *and* P *is a polynomial, then*

$$P(xD)f \overset{ops}{\longrightarrow} \{P(n)a_n\}_{n \geq 0}.$$

Example 1.

Find a closed formula for the sum of the series $\sum_{n \geq 0} (n^2 + 4n + 5)/n!$.
According to the rule, the answer is the value at $x = 1$ of the series

$$\{(xD)^2 + 4(xD) + 5\}e^x = \{x^2 + x\}e^x + 4xe^x + 5e^x$$
$$= (x^2 + 5x + 5)e^x.$$

Therefore the answer to the question is $11e$.

But we cheated. Did you catch the illegal move? We took our generating function and evaluated it at $x = 1$, didn't we? Such an operation doesn't exist in the ring of formal series. There, series don't have 'values' at particular values of x. The letter x is purely a formal symbol whose powers mark the clothespins on the line.

What *can* be evaluated at a particular numerical value of x is a power series that converges at that x, which is an analytic idea rather than a formal one. The way we make peace with our consciences in such situations, which occur frequently, is this: if, after writing out the recurrence relation and solving it by means of a formal power series generating function, we find that the series so obtained converges to an analytic function inside a certain disk in the complex plane, then the whole derivation that we did formally is actually valid analytically for all complex x in that disk. Therefore we can shift gears and regard the series as a convergent analytic creature if it pleases us to do so. ∎

Example 2.

Find a closed formula for the sum of the squares of the first N positive integers.

To do that, begin with the fact that

$$\sum_{n=0}^N x^n = \frac{x^{N+1} - 1}{x - 1},$$

and notice that if we apply $(xD)^2$ to both sides of this relation and then set $x = 1$, the left side will be the sum of squares that we seek, and the right side will be the answer! Hence

$$\sum_{n=1}^N n^2 = (xD)^2 \left\{ \frac{x^{N+1} - 1}{x - 1} \right\} \Bigg|_{x=1}.$$

After doing the two differentiations and lots of algebra, the answer emerges as

$$\sum_{n=1}^{N} n^2 = \frac{N(N+1)(2N+1)}{6} \qquad (N = 1, 2, \ldots),$$

which you no doubt knew already. Do notice, however, that the generating function machine is capable of doing, quite mechanically, many formidable-looking problems involving sums. ∎

Our third rule will be a restatement of the way that two opsgf's are multiplied.

Rule 3. *If* $f \overset{ops}{\longleftrightarrow} \{h_n\}_0^{\infty}$ *and* $g \overset{ops}{\longleftrightarrow} \{b_n\}_0^{\infty}$, *then*

$$fg \overset{ops}{\longleftrightarrow} \left\{\sum_{r=0}^{n} a_r b_{n-r}\right\}_{n=0}^{\infty}. \qquad (2.2.3)$$

Now consider the product of more than two series. For instance, in the case of three series, if f, g, h are the series, and if they generate sequences **a**, **b** and **c**, respectively, then a brief computation shows that fgh generates the sequence

$$\left\{\sum_{r+s+t=n} a_r b_s c_t\right\}_{n=0}^{\infty}. \qquad (2.2.4)$$

A comparison with Rule 3 above will suggest the general formulas that apply to products of any number of power series. One case of this is worth writing down, namely the expressions for the kth power of a series.

Rule 4. *Let* $f \overset{ops}{\longleftrightarrow} \{h_n\}_0^{\infty}$, *and let* k *be a positive integer. Then*

$$f^k \overset{ops}{\longleftrightarrow} \left\{\sum_{n_1+n_2+\cdots+n_k=n} a_{n_1} a_{n_2} \cdots a_{n_k}\right\}_{n=0}^{\infty}. \qquad (2.2.5)$$

Example 3.

Let $f(n, k)$ denote the number of ways that the nonnegative integer n can be written as an ordered sum of k nonnegative integers. Find $f(n, k)$. For instance, $f(4, 2) = 5$ because 4=4+0=3+1=2+2=1+3=0+4.

To find f, consider the power series $1/(1-x)^k$. Since $1/(1-x) \overset{ops}{\longleftrightarrow} \{1\}$, by (2.2.5) we have

$$1/(1-x)^k \overset{ops}{\longleftrightarrow} \{f(n, k)\}_{n=0}^{\infty}.$$

By (1.5.10), $f(n, k) = \binom{n+k-1}{n}$, and we are finished. ∎

Next consider the effect of multiplying a power series by $1/(1-x)$. Suppose $f \overset{ops}{\longleftrightarrow} \{h_n\}_0^{\infty}$. Then what sequence does $f(x)/(1-x)$ generate?

To find out, we have

$$\frac{f(x)}{(1-x)} = (a_0 + a_1 x + a_2 x^2 + \cdots)(1 + x + x^2 + \cdots)$$

$$= a_0 + (a_0 + a_1)x + (a_0 + a_1 + a_2)x^2$$
$$+ (a_0 + a_1 + a_2 + a_3)x^3 + \cdots$$

which clearly leads us to:

Rule 5. *If $f \overset{ops}{\longrightarrow} \{h_n\}_0^\infty$ then*

$$\frac{f}{(1-x)} \overset{ops}{\longrightarrow} \left\{ \sum_{j=0}^{n} a_j \right\}_{n \geq 0}.$$

That is, the effect of dividing an opsgf by $(1-x)$ is to replace the sequence that is generated by the sequence of its partial sums.

Example 4.

Here is another derivation of the formula for the sum of the squares of the first n whole numbers. Since $1/(1-x) \overset{ops}{\longrightarrow} \{1\}_{n \geq 0}$, we have by Rule 2, $(xD)^2(1/(1-x)) \overset{ops}{\longrightarrow} \{n^2\}_{n \geq 0}$, and by Rule 5,

$$\frac{1}{1-x}(xD)^2 \frac{1}{1-x} \overset{ops}{\longrightarrow} \left\{ \sum_{j=0}^{n} j^2 \right\}_{n \geq 0}.$$

That is, the sum of the squares of the first n positive integers is the coefficient of x^n in the series

$$\frac{1}{1-x}(xD)^2 \frac{1}{1-x} = \frac{x(1+x)}{(1-x)^4}.$$

However, by (1.5.10) with $m = 3$,

$$[x^n]\left(\frac{1}{(1-x)^4} \right) = \binom{n+3}{3}.$$

Hence, by (1.2.7),

$$[x^n]\frac{x(1+x)}{(1-x)^4} = \binom{n+2}{3} + \binom{n+1}{3}$$
$$= \frac{n(n+1)(2n+1)}{6},$$

so this must be the sum of the squares of the first n positive integers. ∎

Example 5.

The *harmonic numbers* $\{H_n\}_1^\infty$ are defined by

$$H_n = 1 + \frac{1}{2} + \frac{1}{3} + \cdots + \frac{1}{n} \qquad (n \geq 1).$$

How can we find their ops generating function? By Rule 5, that function is $1/(1-x)$ times the opsgf of the sequence $\{1/n\}_1^\infty$ of reciprocals of the positive integers. So what is $f = \sum_{n \geq 1} x^n/n$? Well, its derivative is $1/(1-x)$, so it must be $-\log(1-x)$. That means that the opsgf of the harmonic numbers is

$$\sum_{n=1}^\infty H_n x^n = \frac{1}{1-x} \log\left(\frac{1}{1-x}\right).$$

∎

Example 6.

Prove that the Fibonacci numbers satisfy

$$F_0 + F_1 + F_2 + \cdots + F_n = F_{n+2} - 1 \qquad (n \geq 0).$$

By Rule 5, the opsgf of the sequence on the left side is $F/(1-x)$, where F is the opsgf of the Fibonacci numbers, which we found in section 1.3 to be $x/(1-x-x^2)$. By Rule 1, the opsgf of the sequence on the right hand side is

$$\frac{F-x}{x^2} - \frac{1}{1-x},$$

and it is the work of just a moment to check that these are equal. ∎

Example 7.

By a *fountain* of coins we mean an arrangement of n coins in rows such that the coins in the first row form a single contiguous block, and that in all higher rows each coin touches exactly two coins from the row beneath it. If the first row contains k coins, we will speak of an (n, k)-fountain. In Fig. 2.1 we show a $(28, 12)$ fountain.

Among all possible fountains we distinguish a special type: those in which *every* row consists of just a single contiguous block of coins. Let's call these *block fountains*.

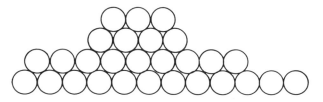

Fig. 2.1: A $(28, 12)$ fountain

The question here is this: how many block fountains have a first row that consists of exactly k coins?

Let $f(k)$ be that number, for $k = 0, 1, 2, \ldots$ If we strip off the first row from such a block fountain, then we are looking at another block fountain that has k fewer coins in it. Conversely, if we wish to form all possible block fountains whose first row has k coins, then begin by laying down that row. Then choose a number j, $0 \leq j \leq k - 1$. Above the row of k coins we will place a block fountain whose first row has j coins. If $j = 0$ there is just one way to do that. Otherwise there are $k - j$ ways to do it, depending on how far in we indent the row of j over the row of k coins. It follows that $f(0) = 1$ and

$$f(k) = \sum_{j=1}^{k} (k - j) f(j) + 1 \qquad (k = 1, 2, \ldots). \qquad (2.2.6)$$

Define the opsgf $F(x) = \sum_{j \geq 0} f(j) x^j$. The appearance, under the summation sign in (2.2.6), of a function of $k - j$ times a function of j should trigger a reflex reaction that Rule 3, above, applies, and that the product of two ordinary power series generating functions is involved. The two series in question are the opsgf's of the integers $\{j\}_1^\infty$ and of the unknowns $\{f(j)\}_1^\infty$, respectively.

However the former opsgf is $x/(1-x)^2$, and the latter is $F(x) - 1$. Hence, after multiplying equation (2.2.6) by x^k and summing over $k \geq 1$ we obtain

$$F(x) - 1 = \frac{x}{(1-x)^2}(F(x) - 1) + \frac{x}{1-x},$$

and therefore

$$F(x) = \frac{1 - 2x}{1 - 3x + x^2}. \qquad (2.2.7)$$

The sequence $\{f(k)\}_0^\infty$ begins with 1, 1, 2, 5, 13, 34, 89, ... If these numbers look suspiciously like Fibonacci numbers, then see exercise 19. ∎

2.3 The calculus of formal exponential generating functions

In this section we will investigate the analogues of the rules in the preceding section, which applied to *ordinary* power series, in the case of *exponential* generating functions.

Definition. *The symbol $f \overset{egf}{\longleftrightarrow} \{h_n\}_0^\infty$ means that the series f is the exponential generating function of the sequence $\{h_n\}_0^\infty$, i.e., that*

$$f = \sum_{n \geq 0} \frac{a_n}{n!} x^n.$$

Let's ask the same questions as in the previous section. Suppose $f \xrightarrow{egf} \{h_n\}_0^\infty$. Then what is the egf of the sequence $\{a_{n+1}\}_0^\infty$? We claim that the answer is f', because

$$f' = \sum_{n=1}^\infty \frac{n a_n x^{n-1}}{n!}$$

$$= \sum_{n=1}^\infty \frac{a_n x^{n-1}}{(n-1)!}$$

$$= \sum_{n=0}^\infty \frac{a_{n+1} x^n}{n!}$$

which is exactly equivalent to the assertion that $f' \xrightarrow{egf} \{a_{n+1}\}_0^\infty$.

Hence the situation with exponential generating functions is just a trifle simpler, in this respect, than the corresponding situation for ordinary power series. Displacement of the subscript by 1 unit in a sequence is equivalent to action of the operator D on the generating function, as opposed to the operator $(f(x) - f(0))/x$, in the case of opsgf's. Therefore we have, by induction:

Rule 1'. If $f \xrightarrow{egf} \{h_n\}_0^\infty$ then, for integer $h \geq 0$,

$$\{a_{n+h}\}_0^\infty \xrightarrow{egf} D^h f. \tag{2.3.1}$$

The reader is invited to compare this Rule 1' with Rule 1, stated above.

Example 1.

To get a hint of the strength of this point of view in problem solving, let's find the egf of the Fibonacci numbers. Now, with just a glance at the recurrence

$$F_{n+2} = F_{n+1} + F_n \qquad (n \geq 0)$$

we see from Rule 1' that the egf satisfies the differential equation

$$f'' = f' + f.$$

At the corresponding stage in the solution for the ops version of this problem, we had an equation to solve for f that did not involve any derivatives. We solved it and then had to deal with a partial fraction expansion in order to find an exact formula for the Fibonacci numbers. In this version, we solve the differential equation, getting

$$f(x) = c_1 e^{r_+ x} + c_2 e^{r_- x} \qquad (r_\pm = (1 \pm \sqrt{5})/2)$$

where c_1 and c_2 are to be determined by the initial conditions (which haven't been used yet!) $f(0) = 0$; $f'(0) = 1$. After applying these two conditions, we find that $c_1 = 1/\sqrt{5}$ and $c_2 = -1/\sqrt{5}$, from which the egf of the Fibonacci sequence is

$$f = (e^{r+x} - e^{r-x})/\sqrt{5}. \qquad (2.3.2)$$

Now it's easier to get the exact formula, because no partial fraction expansion is necessary. Just apply the operator $[x^n/n!]$ to both sides of (2.3.2) and the formula (1.3.3) materializes.

To compare, then, the ops method in this case involves an easier functional equation to solve for the generating function: it's algebraic instead of differential. The egf method involves an easier trip from there to the exact formula, because the partial fraction expansion is unnecessary. Both methods work, which is, after all, the primary *desideratum*. ■

To continue, we discuss next the analogue of Rule 2 for egf's, and that one is easy: it's the same. That is, multiplication of the members of a sequence by a polynomial in n is equivalent to acting on the egf with the same polynomial in the operator xD, and we have:

Rule 2′. *If $f \overset{egf}{\longrightarrow} \{h_n\}_0^\infty$, and P is a given polynomial, then*

$$P(xD)f \overset{egf}{\longrightarrow} \{P(n)a_n\}_{n\geq 0}.$$

Next let's think about the analogue of Rule 3, i.e., about what happens to sequences when their egf's are multiplied together. Precisely, suppose $f \overset{egf}{\longrightarrow} \{h_n\}_0^\infty$ and $g \overset{egf}{\longrightarrow} \{b_n\}_0^\infty$. The question is, of what sequence is fg the egf?

This turns out to have a pretty, and uncommonly useful, answer. To find it, we carry out the multiplication fg and try to identify the coefficient of $x^n/n!$. We obtain

$$
\begin{aligned}
fg &= \left\{ \sum_{r=0}^\infty \frac{a_r x^r}{r!} \right\}\left\{ \sum_{s=0}^\infty \frac{b_s x^s}{s!} \right\} \\
&= \sum_{r,s\geq 0} \frac{a_r b_s}{r!s!} x^{r+s} \\
&= \sum_{n\geq 0} x^n \left\{ \sum_{r+s=n} \frac{a_r b_s}{r!s!} \right\}.
\end{aligned}
$$

The coefficient of $x^n/n!$ is evidently

$$
\begin{aligned}
\left[\frac{x^n}{n!}\right](fg) &= \sum_{r+s=n} \frac{n! a_r b_s}{r!s!} \\
&= \sum_r \binom{n}{r} a_r b_{n-r}.
\end{aligned}
$$

We state this result as:

Rule 3'. *If $f \overset{egf}{\longrightarrow} \{h_n\}_0^\infty$ and $g \overset{egf}{\longrightarrow} \{b_n\}_0^\infty$, then fg generates the sequence*

$$\left\{ \sum_r \binom{n}{r} a_r b_{n-r} \right\}_{n=0}^\infty . \tag{2.3.3}$$

This rule should be contrasted with Rule 3, the corresponding rule for multiplication of opsgf's, the result of which is to generate the sequence

$$\left\{ \sum_r a_r b_{n-r} \right\}_{n=0}^\infty . \tag{2.3.4}$$

We remarked earlier that the convolution of sequences that is shown in (2.3.4) is useful in counting problems where structures of size n are obtained by stitching together structures of sizes r and $n-r$ in all possible ways. Correspondingly, the convolution (2.3.3) is useful in combinatorial situations where we not only stitch together two such structures, but we also *relabel* the structures. For then, roughly speaking, there are $\binom{n}{r}$ ways to choose the new labels of the elements of the structure of size r, as well as a_r ways to choose that structure and b_{n-r} ways to choose the other one.

Since this no doubt all seems to be very abstract, let's try to make it concrete with a few examples.

Example 2.

In (1.6.13) we found the recurrence formula for the Bell numbers, which we may write in the form

$$b(n+1) = \sum_k \binom{n}{k} b(k) \qquad (n \geq 0; b(0) = 1). \tag{2.3.5}$$

We will now apply the methods of this section to find the egf of the Bell numbers. This will give an independent proof of Theorem 1.6.1, since (2.3.5) can be derived directly, as described in exercise 7 of chapter 1.

Let B be the required egf. The egf of the left side of (2.3.5) is, by Rule 1', B'. If we compare the right side of (2.3.5) with (2.3.3) we see that the egf of the sequence on the right of (2.3.5) is the product of B and the egf of the sequence whose entries are all 1's. This latter egf is evidently e^x, and so we have

$$B' = e^x B$$

as the equation that we must solve in order to find the unknown egf. But obviously the solution is $B = c \exp(e^x)$, and since $B(0) = 1$, we must have $c = e^{-1}$, from which $B(x) = \exp(e^x - 1)$, completing the re-proof of Theorem 1.6.1. ∎

Example 3.

In order to highlight the strengths of ordinary vs. exponential generating functions, let's do a problem where the form of the convolution of sequences that occurs suggests the ops form of generating function. We will count the ways of arranging n pairs of parentheses, each pair consisting of a left and a right parenthesis, into a legal string. A legal string of parentheses is one with the property that, as we scan the string from left to right we never will have seen more right parentheses than left.

There are exactly 5 legal strings of 3 pairs of parentheses, namely:

$$((())); \ (()()); \ (())(); \ ()()(); \ ()(()). \qquad (2.3.6)$$

Let $f(n)$ be the number of legal strings of n pairs of parentheses ($f(0) = 1$), for $n \geq 0$.

With each legal string we associate a unique nonnegative integer k, as follows: as we scan the string from left to right, certainly after we have seen all n pairs of parentheses, the number of lefts will equal the number of rights. However, these two numbers may be equal even earlier than that. In the last string in (2.3.6), for instance, after just $k = 1$ pairs have been scanned, we find that all parentheses that have been opened have also been closed. In general, for any legal string, the integer k that we associate with it is the *smallest* positive integer such that the first $2k$ characters of the string do themselves form a legal string. The values of k that are associated with each of the strings in (2.3.6) are 3, 3, 2, 1, 1. We will say that a legal string of $2n$ parentheses is *primitive* if it has $k = n$. The first two strings in (2.3.6) are primitive.

How many legal strings of $2n$ parentheses will have a given value of k? Let w be such a string. The first $2k$ characters of w are a primitive string, and the last $2n - 2k$ characters of w are an arbitrary legal string. There are exactly $f(n-k)$ ways to choose the last $2n - 2k$ characters, but in how many ways can we choose the first $2k$? That is, how many *primitive* strings of length $2k$ are there?

Lemma 2.3.1. *If $k \geq 1$ and $g(k)$ is the number of primitive legal strings, and $f(k)$ is the number of all legal strings of $2k$ parentheses, then*

$$g(k) = f(k - 1).$$

Proof. Given any legal string of $k-1$ pairs of parentheses, make a primitive one of length $2k$ by adding an initial left parenthesis and a terminal right parenthesis to it. Conversely, given a primitive string of length $2k$, if its initial left and terminal right parentheses are deleted, what remains is an arbitrary legal string of length $2k - 2$. Hence there are as many primitive strings of length $2k$ as there are all legal strings of length $2k - 2$, i.e., there are $f(k - 1)$ of them. ∎

Hence the number of legal strings of length $2n$ that have a given value of k is $f(k-1)f(n-k)$. Since every legal string has a unique value of k, it must be that

$$f(n) = \sum_k f(k-1)f(n-k) \qquad (n \neq 0; f(0) = 1) \qquad (2.3.7)$$

with the convention that $f = 0$ at all negative arguments.

The recurrence easily allows us to compute the values $1, 1, 2, 5, 14, \ldots$ Now let's find a generating function for these numbers. The clue as to which kind of generating function is appropriate comes from the form of the recurrence (2.3.7). The sum on the right is obviously related to the coefficients of the product of two *ordinary* power series generating functions, so that is the species that we will use.

Let $F = \sum_k f(k)x^k$ be the opsgf of $\{f(n)\}_{n \geq 0}$. Then the right side of (2.3.7) is *almost* the coefficient of x^n in the series F^2. What is it *exactly*? It is the coefficient of x^n in the product of the series F and the series $\sum_k f(k-1)x^k$. How is this latter series related to F? It is just xF. Therefore, if we multiply the right side of (2.3.7) by x^n and sum over $n \neq 0$, we get xF^2. If we multiply the left side by x^n and sum over $n \neq 0$, we get $F - 1$. Therefore our unknown generating function satisfies the equation

$$F(x) - 1 = xF(x)^2. \qquad (2.3.8)$$

Here we have a new wrinkle. We are accustomed to going from recurrence relations on a sequence to functional equations that have to be solved for generating functions. In previous examples, those functional equations have either been simple linear equations or differential equations. In (2.3.8) we have a generating function that satisfies a quadratic equation. When we solve it, we get

$$F(x) = \frac{1 \pm \sqrt{1 - 4x}}{2x}.$$

Which sign do we want? If we choose the '+' then the numerator will approach 2 as $x \to 0$, so the ratio will become infinite at 0. But our generating function takes the value 1 at 0, so that can't be right. If we choose the '−' sign, then a dose of L'Hospital's rule shows that we will indeed have $F(0) = 1$. Hence our generating function is

$$F(x) = \frac{1 - \sqrt{1 - 4x}}{2x}. \qquad (2.3.9)$$

This is surely one of the most celebrated generating functions in combinatorics. The numbers $f(n)$ are the *Catalan numbers*, and in (2.5.10) there is an explicit formula for them. For the moment, we declare that this exercise, which was intended to show how the form of a recurrence can guide the choice of generating function, is over. ∎

Example 4.

By a *derangement* of n letters we mean a permutation of them that has no fixed points. Let D_n denote the number of derangements of n letters, and let $D(x) \overset{egf}{\longleftrightarrow} \{D_n\}_0^\infty$. We will find a recurrence for the sequence, then $D(x)$, then an explicit formula for the members of the sequence.

The number of permutations of n letters that have a particular set of $k \leq n$ letters as their set of fixed points is clearly D_{n-k}. There are $\binom{n}{k}$ ways to choose the set of k fixed points, and so there are exactly $\binom{n}{k} D_{n-k}$ permutations of n letters that have exactly k fixed points. Since every permutation has *some* set of fixed points, it must be that

$$n! = \sum_k \binom{n}{k} D_{n-k} \qquad (n \geq 0).$$

If we take the egf of both sides we get, by Rule $3'$,

$$\frac{1}{1-x} = e^x D(x)$$

(see how easy that was?), from which $D(x) = e^{-x}/(1-x)$. Next, by Rule 5, if we take $[x^n]$ on both sides, we find that

$$\frac{D_n}{n!} = 1 - 1 + \frac{1}{2!} - \frac{1}{3!} + \cdots + (-1)^n \frac{1}{n!},$$

and we are finished. ∎

Just as in the case of ordinary power series generating functions, pleasant and useful things happen when we consider products of more than two exponential generating functions. For instance, if we multiply three of them, f, g, and h, which generate $\mathbf{a}$, $\mathbf{b}$, and $\mathbf{c}$, respectively, then we find that

$$fgh \overset{egf}{\longleftrightarrow} \left\{ \sum_{r+s+t=n} \frac{n!}{r!s!t!} a_r b_s c_t \right\}_{n=0}^\infty, \qquad (2.3.10)$$

and therefore such operations can be expected to be helpful in dealing with sums that involve multinomial coefficients.

If $f \overset{egf}{\longleftrightarrow} \{h_n\}_0^\infty$ then

$$f^k \overset{egf}{\longleftrightarrow} \left\{ \sum_{r_1+\cdots+r_k=n} \frac{n!}{r_1! r_2! \cdots r_k!} a_{r_1} a_{r_2} \cdots a_{r_k} \right\}_{n=0}^\infty. \qquad (2.3.11)$$

2.4 Power series, analytic theory

The *formal* theory of power series shows us that we can manipulate recurrences and solve functional equations, such as differential equations, for power series without necessarily worrying about whether the resulting series converge. If they do converge though, and they represent functions, that's a big advantage, for then we may be in a position to find analytic information about the recurrence relation that might not otherwise be easily obtainable.

In this section we will review the basic analytic properties of power series and their coefficient sequences.

First, suppose we are given a power series

$$f = \sum_{n \geq 0} a_n z^n,$$

where we now use the letter z to encourage thinking about complex variables. Question: for exactly what set of complex values of z does the series f converge? We want to give a fairly complete answer to this question, and express it in terms of the coefficient sequence $\{h_n\}_0^\infty$.

Theorem 2.4.1. *There exists a number R, $0 \leq R \leq +\infty$, called the radius of convergence of the series f, such that the series converges for all values of z with $|z| < R$ and diverges for all z such that $|z| > R$. The number R is expressed in terms of the sequence $\{h_n\}_0^\infty$ of coefficients of the series by means of*

$$R = \frac{1}{\limsup_{n \to \infty} |a_n|^{1/n}} \qquad (1/0 = \infty; 1/\infty = 0). \tag{2.4.1}$$

Before proving the theorem, we recall the definition of the *limit superior* of a sequence. Let $\{x_n\}_0^\infty$ be a sequence of real numbers, and let L be a real number (possibly $= \pm\infty$).

Definition. *We say that L is the limit superior ('upper limit') of the sequence $\{x_n\}$ if*

(a) *L is finite and*

(i) *for every $\epsilon > 0$ all but finitely many members of the sequence satisfy $x_n < L + \epsilon$, and*

(ii) *for every $\epsilon > 0$, infinitely many members of the sequence satisfy $x_n > L - \epsilon$, or*

(b) *$L = +\infty$ and for every $M > 0$, there is an n such that $x_n > M$, or*

(c) *$L = -\infty$ and for every x, there are only finitely many n such that $x_n > x$.*

If L is the limit superior of the sequence $\{x_n\}_0^\infty$, then we write $L = \limsup_{n \to \infty}\{x_n\}$, or perhaps just $L = \limsup\{x_n\}$, if the context is clear enough.

The limit superior has the following properties:
- Every sequence of real numbers has one and only one limit superior in the extended real number system (i.e., including $\pm\infty$).
- If a sequence has a *limit* L, then L is also the limit superior of the sequence.
- If S is the set of cluster points of the sequence $\{x_n\}_0^\infty$, then $\limsup\{x_n\}$ is the least upper bound of the numbers in S.

Proof of theorem 2.4.1. Let R be the number shown in (2.4.1), and suppose first that $0 < R < \infty$. Choose z such that $|z| < R$. We will show that the series converges at z.

For the given z, we can find $\epsilon > 0$ such that

$$|z| < \frac{R}{1 + \epsilon R}.$$

Now, by the definition of the $\limsup$, there exists N such that for all $n > N$ we have

$$|a_n|^{1/n} < \frac{1}{R} + \epsilon.$$

Hence, for these same n,

$$|a_n||z|^n < \left\{|z|(\frac{1}{R} + \epsilon)\right\}^n.$$

Let α denote the number in the curly brace. Then by our choice of ϵ, we have $\alpha < 1$. Hence the series $\sum a_n z^n$ converges absolutely, by comparison with the terms of a convergent geometric series. Therefore our series converges absolutely at z, and hence it does so for all $|z| < R$.

Next we claim the series diverges if $|z| > R$. Indeed, we will show that for such z, the sequence of terms of the series does not approach zero. Since $|z| > R$, we can choose $\epsilon > 0$ such that if $\theta = |(z/R) - \epsilon z|$, then $\theta > 1$. By definition of the $\limsup$, for infinitely many values of n we have $|a_n|^{1/n} > (1/R) - \epsilon$. Hence, for those values of n,

$$|a_n z^n| > \left|\left(\frac{1}{R} - \epsilon\right) z\right|^n = \theta^n$$

which increases without bound since $\theta > 1$. Hence, that subsequence of terms of the power series does not approach zero and the series diverges. This completes the proof of the theorem in the case that $0 < R < \infty$. The cases where $R = 0$ or $R = +\infty$ are similar, and are left to the reader. ∎

Theorem 2.4.2. *Suppose the power series $\sum a_n z^n$ converges for all z in $|z| < R$, and let $f(z)$ denote its sum. Then $f(z)$ is an analytic function in*

$|z| < R$. *If furthermore the series diverges for $|z| > R$, then the function $f(z)$ must have at least one singularity on the circle of convergence $|z| = R$.*

In other words: *a power series keeps on converging until something stops it, namely a singularity of the function that is being represented.*
Proof. If f has no singularity on its circle of convergence $|z| = R$, then about each point of that circle we can draw an open disk in which f remains analytic. By the Heine-Borel theorem, a finite number of these disks cover the circle $|z| = R$, and therefore f must remain analytic in some larger disk $|z| < R + \epsilon$. By Cauchy's inequality, the Taylor coefficients of the series for f satisfy $|a_n| \le M/(R+\epsilon)^n$, for all n, and so the series must converge in a larger disk, a contradiction. ■

Example 1.

The series $\sum z^n$ converges if $|z| < 1$ and diverges if $|z| > 1$. Hence the function that is represented must have a singularity somewhere on the circle $|z| = 1$. That function is $1/(1 - z)$, and sure enough it has a singularity at $z = 1$. ■

Example 2.

Take the function $f(z) = 1/(2 - e^z)$. Suppose we expand $f(z)$ in a power series about $z = 0$. What will be the radius of convergence of the series?

According to the theorem, the series will converge in the largest disk $|z| < R$ in which f is analytic. The function fails to be analytic only at the points z where $e^z = 2$. Those points are of the form $z = \log 2 + 2k\pi i$, for integer k, and the nearest one to the origin is $\log 2$. Therefore $f(z)$ is analytic in the disk $|z| < \log 2$ and in no larger disk. Hence the radius of convergence of the series will be $R = \log 2$. ■

Remember that, if $f(z)$ is given, *the best way to find the radius of convergence of its power series expansion about the origin may well be to look for its singularity that is nearest to the origin.*

Example 3.

Take the function $f(z) = z/(e^z - 1)$ $(f(0) = 1)$. Estimate the size of the coefficients of its power series about the origin directly from the analyticity properties of the function.

This is where things start getting more interesting. This $f(z)$ is analytic except possibly at points z where $e^z = 1$, i.e., except possibly at the points $z = 2k\pi i$ for integer k. The nearest of these to the origin is the origin itself ($k = 0$). However, f is not singular at $z = 0$ because even though the denominator of f is 0 there, the numerator is also, and L'Hospital's rule, or whatever, reveals that the value $f(0) = 1$ removes the singularity. Hence the singularity of this function that is nearest to the origin is at $z = 2\pi i$.

The power series

$$\frac{z}{e^z - 1} = \sum_{n=0}^{\infty} a_n z^n$$

therefore has radius of convergence $R = 2\pi$.

The problem asks for estimates of the sizes of the coefficients $\{h_n\}_0^{\infty}$. But since the radius of convergence is 2π, we have, from theorem 2.4.1,

$$\limsup |a_n|^{1/n} = \frac{1}{2\pi}.$$

It follows that, first of all, for all sufficiently large values of n we have

$$|a_n|^{1/n} < \frac{1}{2\pi} + \epsilon,$$

and for infinitely many values of n we have

$$|a_n|^{1/n} > \frac{1}{2\pi} - \epsilon.$$

Therefore, what we find out about the coefficients is that for each $\epsilon > 0$, there exists N such that

$$|a_n| < \left(\frac{1}{2\pi} + \epsilon\right)^n \qquad (n > N)$$

and further, for infinitely many values of n,

$$|a_n| > \left(\frac{1}{2\pi} - \epsilon\right)^n.$$

Therefore the coefficients of this series decrease to zero exponentially fast, at roughly the rate of $1/(2\pi)^n$, for large n. This is quite a lot to have found out about the sizes of the coefficients without having calculated any of them! ∎

We state, for future reference, a general proposition that summarizes what we learned in this example.

Theorem 2.4.3. *Let $f(z) = \sum a_n z^n$ be analytic in some region containing the origin, let a singularity of $f(z)$ of smallest modulus be at a point $z_0 \neq 0$, and let $\epsilon > 0$ be given. Then there exists N such that for all $n > N$ we have*

$$|a_n| < \left(\frac{1}{|z_0|} + \epsilon\right)^n.$$

Further, for infinitely many n we have

$$|a_n| > \left(\frac{1}{|z_0|} - \epsilon\right)^n.$$

In chapter 5 we will learn how to make much more precise estimates of the sizes of the coefficients of power series based on the analyticity, or lack thereof, of the function that is represented by the series. For instance, the method of Darboux (Theorem 5.3.1) is a powerful technique for asymptotic analysis of coefficient sequences of generating functions. The existence of such methods is an excellent reason why we should be knowledgeable about the analytic, as well as the formal, side of the subject of generating functions.

Another path to the asymptotic analysis of coefficient sequences flows from Cauchy's formula

$$a_n = \frac{1}{2\pi i} \int \frac{f(z)dz}{z^{n+1}} \qquad (n = 0, 1, 2, \ldots) \qquad (2.4.2)$$

that expresses the nth coefficient of the Taylor's series expansion $f(z) = \sum a_n z^n$ as a contour integral involving the function f. The contour can be any simple, closed curve that encloses the origin and that lies entirely inside a region in which f is analytic.

One has immediately, from (2.4.2), Cauchy's inequality, which states that

$$|a_n| \leq \frac{M(r)}{r^n},$$

and which holds for all $n \geq 0$ and all $0 < r < R$, where R is the radius of convergence of the series, and

$$M(r) = \max_{|z| \leq r} |f(z)| = \max_{|z| = r} |f(z)|. \qquad (2.4.3)$$

Just as the analysis of Example 3 above is refined by the method of Darboux to a much more precise method of estimating the growth of coefficient sequences, so is Cauchy's inequality refined by the method of Hayman (Theorem 5.4.1) to another very precise tool for the same purpose.

If a power series actually converges to a function, then we can use roots of unity to pick out a progression of terms from a series. For instance, how can we select just the *even* powers out of a power series? If the series represents a function f, then, as is well known, $(f(x) + f(-x))/2$ has just the terms that involve even powers of x from the series for $f(x)$, and $(f(x) - f(-x))/2$ has just the odd ones.

But suppose, instead of wanting to keep every second term of the series, we want to keep only every third term? For instance, what function do we get if we take the exponential series and keep just the terms where the powers of x are multiples of 3? In other words, who is

$$g(x) = \sum_{n \geq 0} \frac{x^{3n}}{(3n)!} \ ? \qquad (2.4.4)$$

Well, what makes the $(f(x) + f(-x))/2$ thing work is that the two *square roots of unity*, namely ± 1, have the property that

$$\frac{1^n + (-1)^n}{2} = \begin{cases} 1, & \text{if } n \text{ is even}; \\ 0, & \text{if } n \text{ is odd}. \end{cases}$$

Now here is a correspondingly helpful property of the three *cube roots* of unity $1, \omega_1, \omega_2$:

$$\frac{(1^n + \omega_1^n + \omega_2^n)}{3} = \begin{cases} 1, & \text{if } 3 \backslash n; \\ 0, & \text{else}. \end{cases} \tag{2.4.5}$$

Since that is the case, we have, for any convergent power series $f = \sum_r a_r x^r$,

$$\frac{f(x) + f(\omega_1 x) + f(\omega_2 x)}{3} = \sum_r a_{3r} x^{3r}. \tag{2.4.6}$$

Since $\omega_1 = e^{(2\pi i)/3}$ and $\omega_2 = e^{(4\pi i)/3}$, we can unmask the mystery function $g(x)$ in (2.4.4) as

$$\begin{aligned} g(x) &= \frac{1}{3} \left(e^x + e^{\omega_1 x} + e^{\omega_2 x} \right) \\ &= \frac{1}{3} \left(e^x + 2e^{-x/2} \cos\left(\frac{\sqrt{3}x}{2}\right) \right). \end{aligned} \tag{2.4.7}$$

Example 4.

For fixed n, find

$$\lambda_n = \sum_k (-1)^k \binom{n}{3k}.$$

We could do this one if we knew the function

$$f(x) = \sum_k \binom{n}{3k} x^{3k},$$

because $\lambda_n = f(-1)$. But $f(x)$ picks out every third term from the series $F(x) = (1 + x)^n$, and so

$$\begin{aligned} f(x) &= (F(x) + F(\omega_1 x) + F(\omega_2 x))/3 \\ &= \{(1 + x)^n + (1 + \omega_1 x)^n + (1 + \omega_2 x)^n\}/3. \end{aligned}$$

Thus the numbers that we are asked to find are, for $n > 0$,

$$\begin{aligned} \lambda_n = f(-1) &= \{(1 - \omega_1)^n + (1 - \omega_2)^n\}/3 \\ &= \frac{1}{3} \left\{ \left(\frac{3 - \sqrt{3}i}{2}\right)^n + \left(\frac{3 + \sqrt{3}i}{2}\right)^n \right\} \tag{2.4.8} \\ &= 2 \cdot 3^{(n/2-1)} \cos\left(\frac{n\pi}{6}\right). \end{aligned}$$

The first few values of the $\{\lambda_n\}_{n\geq 0}$ are 1, 1, 1, 0, -3, -9, -18, ...

To complete this example we want to prove the helpful property (2.4.5) of the cube roots of unity. But for every $r > 1$, the rth roots of unity do the same sort of thing, namely

$$\frac{1}{r} \sum_{\omega^r=1} \omega^n = \begin{cases} 1 & \text{if } r \backslash n \\ 0 & \text{else.} \end{cases} \tag{2.4.9}$$

Indeed, the left side is

$$\frac{1}{r} \sum_{j=0}^{r-1} e^{(2\pi i j n)/r},$$

which is a finite geometric series whose sum is easy to find, and is as stated in (2.4.9). So, with more or less difficulty, it is always possible to select a subset of the terms of a convergent series in which the exponents form an arithmetic progression. See exercise 25. ∎

2.5 Some useful power series

Generatingfunctionologists need reference lists of known power series and other series that occur frequently in applications of the theory. Here is such a list. For each series we show the series and its sum. The radius of the largest open disk, centered at the origin, in which convergence takes place will be, of course, the modulus of the singularity of the function that is nearest to the origin. Considering the relatively simple forms of the functions, the locations of those singularities will be sufficiently obvious that the radii of convergence are not explicitly shown in the table below.

$$\frac{1}{1-x} = \sum_{n\geq 0} x^n \tag{2.5.1}$$

$$\log \frac{1}{1-x} = \sum_{n\geq 1} \frac{x^n}{n} \tag{2.5.2}$$

$$e^x = \sum_{n\geq 0} \frac{x^n}{n!} \tag{2.5.3}$$

$$\sin x = \sum_{n\geq 0} (-1)^n \frac{x^{2n+1}}{(2n+1)!} \tag{2.5.4}$$

$$\cos x = \sum_{n\geq 0} (-1)^n \frac{x^{2n}}{(2n)!} \tag{2.5.5}$$

$$(1 + x)^\alpha = \sum_k \binom{\alpha}{k} x^k \qquad (2.5.6)$$

$$\frac{1}{(1 - x)^{k+1}} = \sum_n \binom{n + k}{n} x^n \qquad (2.5.7)$$

$$\frac{x}{e^x - 1} = \sum_{n \geq 0} \frac{B_n x^n}{n!} \qquad (2.5.8)$$

$$\tan^{-1} x = \sum_{n \geq 0} (-1)^n \frac{x^{2n+1}}{2n + 1} \qquad (2.5.9)$$

$$\frac{1}{2x}(1 - \sqrt{1 - 4x}) = \sum_n \frac{1}{n + 1} \binom{2n}{n} x^n \qquad (2.5.10)$$

$$= 1 + x + 2x^2 + 5x^3 + 14x^4 + 42x^5 + 132x^6$$
$$+ 429x^7 + 1430x^8 + 4862x^9 + \cdots$$

$$\frac{1}{\sqrt{1 - 4x}} = \sum_k \binom{2k}{k} x^k \qquad (2.5.11)$$

$$= 1 + 2x + 6x^2 + 20x^3 + 70x^4 + 252x^5 + 924x^6$$
$$+ 3432x^7 + 12870x^8 + 48620x^9 + \cdots$$

$$x \cot x = \sum_{k \geq 0} \frac{(-4)^k B_{2k}}{(2k)!} x^{2k} \qquad (2.5.12)$$

$$= 1 - \frac{x^2}{3} - \frac{x^4}{45} - \frac{2\,x^6}{945} - \frac{x^8}{4725} - \frac{2\,x^{10}}{93555} - \cdots$$

$$\tan x = \sum_{r \geq 1} (-1)^{r-1} \frac{2^{2r}(2^{2r} - 1)B_{2r}}{(2r)!} x^{2r-1} \qquad (2.5.13)$$

$$= x + \frac{x^3}{3} + \frac{2\,x^5}{15} + \frac{17\,x^7}{315} + \frac{62\,x^9}{2835} + \frac{1382\,x^{11}}{155925} + \cdots$$
$$+ \frac{21844\,x^{13}}{6081075} + \frac{929569\,x^{15}}{638512875} + \cdots$$

$$\frac{x}{\sin x} = \sum_{r \geq 0} (-1)^{r-1} \frac{(4^r - 2)B_{2r}}{(2r)!} x^{2r}$$

$$= 1 + \frac{x^2}{6} + \frac{7x^4}{360} + \frac{31x^6}{15120} + \cdots \tag{2.5.14}$$

$$\frac{1}{\sqrt{1-4x}} \left(\frac{1 - \sqrt{1-4x}}{2x} \right)^k = \sum_n \binom{2n+k}{n} x^n \tag{2.5.15}$$

$$\left(\frac{1 - \sqrt{1-4x}}{2x} \right)^k = \sum_{n \geq 0} \frac{k(2n+k-1)!}{n!(n+k)!} x^n \qquad (k \geq 1) \tag{2.5.16}$$

$$\sin^{-1}(x) = x + \frac{1}{2}\frac{x^3}{3} + \frac{1 \cdot 3}{2 \cdot 4}\frac{x^5}{5} + \frac{1 \cdot 3 \cdot 5}{2 \cdot 4 \cdot 6}\frac{x^7}{7} + \cdots \tag{2.5.17}$$

$$e^x \sin x = \sum_{n \geq 1} \frac{2^{\frac{n}{2}} \sin \frac{n\pi}{4}}{n!} x^n \tag{2.5.18}$$

$$= x + x^2 + \frac{x^3}{3} - \frac{x^5}{30} - \frac{x^6}{90} - \frac{x^7}{630} + \cdots$$

$$\frac{1}{2} \tan^{-1}(x) \log(1 + x^2) = \sum_{r \geq 1} (-1)^{r-1} H_{2r} \frac{x^{2r+1}}{2r+1} \tag{2.5.19}$$

$$= \frac{x^3}{2} - \frac{5x^5}{12} + \frac{7x^7}{10} - \frac{761x^9}{1260} + \cdots$$

$$\frac{1}{4} \tan^{-1}(x) \log \frac{1+x}{1-x} = \sum_{r \geq 0} \frac{x^{4r+2}}{4r+2} \left(1 - \frac{1}{3} + \frac{1}{5} - \cdots + \frac{1}{4r+1} \right)$$

$$= \frac{x^2}{2} + \frac{13x^6}{90} + \frac{263x^{10}}{3150} + \cdots \tag{2.5.20}$$

$$\frac{1}{2} \left\{ \log \frac{1}{1-x} \right\}^2 = \sum_{r \geq 2} \frac{H_{r-1}}{r} x^r \tag{2.5.21}$$

$$= \frac{x^2}{2} + \frac{x^3}{2} + \frac{11x^4}{24} + \frac{5x^5}{12} + \frac{137x^6}{360} + \frac{7x^7}{20} + \cdots$$

$$\sqrt{\frac{1 - \sqrt{1-x}}{x}} = \sum_{k=0}^{\infty} \frac{(4k)!}{16^k \sqrt{2}(2k)!(2k+1)!} x^k \qquad (2.5.22)$$

$$= \frac{1}{\sqrt{2}}\left(1 + \frac{x}{8} + \frac{7\,x^2}{128} + \frac{33\,x^3}{1024} + \frac{715\,x^4}{32768} + \frac{4199\,x^5}{262144}\right.$$

$$+ \frac{52003\,x^6}{4194304} + \frac{334305\,x^7}{33554432} + \frac{17678835\,x^8}{2147483648}$$

$$+ \frac{119409675\,x^9}{17179869184} + \frac{1641030105\,x^{10}}{274877906944} + \cdots\bigg)$$

$$e^{\arcsin x} = \sum_{k=0}^{\infty} \frac{\prod_{j=0}^{k-1}(4j^2+1)}{(2k)!} x^{2k} + \sum_{k=0}^{\infty} \frac{4^k \prod_{j=1}^{k}(\frac{1}{2} - j + j^2)}{(2k+1)!} x^{2k+1}$$

$$= 1 + x + \frac{x^2}{2} + \frac{x^3}{3} + \frac{5\,x^4}{24} + \frac{x^5}{6} + \frac{17\,x^6}{144} + \frac{13\,x^7}{126}$$

$$+ \frac{629\,x^8}{8064} + \frac{325\,x^9}{4536} + \frac{8177\,x^{10}}{145152} + \cdots \qquad (2.5.23)$$

$$\left(\frac{\arcsin x}{x}\right)^2 = \sum_{k=0}^{\infty} \frac{4^k k!^2}{(k+1)(2k+1)!} x^{2k} \qquad (2.5.24)$$

$$= 1 + \frac{x^2}{3} + \frac{8\,x^4}{45} + \frac{4\,x^6}{35} + \frac{128\,x^8}{1575} + \frac{128\,x^{10}}{2079} + \cdots$$

$$(x + \sqrt{1+x^2})^a = \sum_{k=0}^{\infty} \frac{2^k \cdot (\frac{a}{2} - \frac{k}{2} + 1)^{\overline{k}}}{(1 + k/a)k!} x^k \qquad (2.5.25)$$

$$= 1 + a\,x + \frac{a^2\,x^2}{2} + \left(\frac{-a}{6} + \frac{a^3}{6}\right) x^3 + \left(\frac{-a^2}{6} + \frac{a^4}{24}\right) x^4$$

$$+ \frac{a\,(9 - 10\,a^2 + a^4)\,x^5}{120} + \frac{a^2\,(64 - 20\,a^2 + a^4)\,x^6}{720}$$

$$+ \frac{a\,(-225 + 259\,a^2 - 35\,a^4 + a^6)\,x^7}{5040}$$

$$+ \frac{a^2\,(-2304 + 784\,a^2 - 56\,a^4 + a^6)\,x^8}{40320} + \cdots$$

In the above, the $\{B_n\}$ are the *Bernoulli numbers*, and they are defined by (2.5.8). The Bernoulli numbers $\{B_n\}_0^{16}$ have the values

$$1,\ -1/2,\ \frac{1}{6},\ 0,\ -\frac{1}{30},\ 0,\ \frac{1}{42},\ 0,\ -\frac{1}{30},\ 0,\ \frac{5}{66},\ 0,\ -\frac{691}{2730},\ 0,\ \frac{7}{6},\ 0,\ -\frac{3617}{510}.$$

The $\{H_n\}$ are the *harmonic numbers* that were defined in section 2.2. The symbol $m^{\overline{k}}$, in (2.5.25), means $m(m+1)\cdots(m+k-1)$. The expansions (2.5.22)-(2.5.25) are taken from [Ko].

2.6 Dirichlet series, formal theory

We have already discussed two slightly different forms of generating functions of sequences, namely the ordinary power series form and the exponential generating function form. We remarked that when, in a particular problem, one has to decide which of these forms to use, the choice is most often dictated by the form of the multiplicative convolution of the two sequences that occurs in the problem. If the form is as in Rule 3' and (2.3.3), then we choose the egf, whereas if it is of the form (2.3.4), the opsgf may well be preferred.

To help highlight the basis for this kind of choice, we will now discuss yet another kind of generating function that matches yet another kind of convolution of two sequences, a kind that also occurs naturally in many problems in combinatorics and number theory.

Definition. *Given a sequence $\{h_n\}_1^\infty$; we say that a formal series*

$$f(s) = \sum_{n=1}^\infty \frac{a_n}{n^s}$$

$$= a_1 + \frac{a_2}{2^s} + \frac{a_3}{3^s} + \frac{a_4}{4^s} + \cdots$$

(2.6.1)

is the Dirichlet series generating function (Dsgf) of the sequence, and we write

$$f(s) \overset{Dir}{\longleftrightarrow} \{h_n\}_1^\infty.$$

The importance of Dirichlet series stems directly from their multiplication rule. Suppose $f(s) \overset{Dir}{\longleftrightarrow} \{h_n\}_1^\infty$ and $g(s) \overset{Dir}{\longleftrightarrow} \{b_n\}_1^\infty$. The question is, what sequence is generated by $f(s)g(s)$?

To find out, consider the product of these series,

$$fg = (a_1 + a_2 2^{-s} + a_3 3^{-s} + \cdots)(b_1 + b_2 2^{-s} + b_3 3^{-s} + \cdots)$$
$$= (a_1 b_1) + (a_1 b_2 + a_2 b_1)2^{-s} + (a_1 b_3 + a_3 b_1)3^{-s}$$
$$+ (a_1 b_4 + a_2 b_2 + a_4 b_1)4^{-s} + \cdots$$

What is the general rule? In the product fg, what is the coefficient of n^{-s}? It is the sum of all products of a's and b's where the product of their subscripts is n, i.e., it is

$$\sum_{rs=n} a_r b_s.$$

Now if $rs = n$ then r and s are divisors of n, so the above sum can also be written as

$$\sum_{d\backslash n} a_d b_{\frac{n}{d}},$$

in which the symbol '$d\backslash n$' is read 'd divides n.' We state this formally as:

Rule 1″. If $f(s) \overset{Dir}{\longleftrightarrow} \{h_n\}_1^\infty$ and $g(s) \overset{Dir}{\longleftrightarrow} \{b_n\}_1^\infty$, then

$$f(s)g(s) \overset{Dir}{\longleftrightarrow} \left\{ \sum_{d\backslash n} a_d b_{\frac{n}{d}} \right\}_{n=1}^\infty. \tag{2.6.2}$$

Let's hasten to say what kind of a problem gives rise to this kind of a convolution of sequences. It is, roughly, a situation in which all objects of size n are obtained by stitching together d objects of size n/d, where d is some divisor of n. Before we get to examples of this sort of thing, since the multiplication is so important, let's look at a few more of its properties.

What happens to the sequence generated if we take the kth power of a Dirichlet series? Let's work it out, as follows:

$$
\begin{aligned}
f(s)^k &= \left(\sum_{n \geq 1} a_n n^{-s} \right)^k \\
&= \sum_{n_1,\ldots,n_k \geq 1} a_{n_1} \cdots a_{n_k} (n_1 n_2 \cdots n_k)^{-s} \\
&= \sum_{n \geq 1} n^{-s} \left\{ \sum_{n_1 \cdots n_k = n} a_{n_1} \cdots a_{n_k} \right\}.
\end{aligned}
$$

This shows:

Rule 2″. If $f(s) \overset{Dir}{\longleftrightarrow} \{h_n\}_1^\infty$ then $f(s)^k \overset{Dir}{\longleftrightarrow}$ a sequence whose nth member is the sum, extended over all ordered factorizations of n into k factors, of the products of the members of the sequence whose subscripts are the factors in that factorization.

What series f generates the sequence of all 1's: $\{1\}_1^\infty$? When we asked that question in the cases of the opsgf and the egf, the answers turned out to be 'famous' functions. For opsgf's it was $1/(1-x)$ and for egf's it was e^x. In the present case, the formal Dirichlet series whose coefficients are all 1's is not related to any simple function of analysis, it is a new creature, and it gets a new name: the *Riemann zeta function*. It is the Dirichlet series

$$\zeta(s) = \sum_{n=1}^{\infty} \frac{1}{n^s}$$

$$= 1^{-s} + 2^{-s} + 3^{-s} + 4^{-s} + \cdots,$$

(2.6.3)

and it is one of the most important functions in analysis.

Now, since $\zeta(s) \xrightarrow{Dir} \{1\}_1^{\infty}$, what sequence does $\zeta^2(s)$ generate? Directly from (2.6.2),

$$[n^{-s}]\zeta^2(s) = \sum_{d \backslash n} 1 \cdot 1 = d(n),$$

where $d(n)$ is the number of divisors of the integer n. The sequence $d(n)$ is quite irregular, and begins with

$$1, 2, 2, 3, 2, 4, 2, 4, 3, 4, 2, \ldots$$

Nevertheless, its Dirichlet series generating function is $\zeta^2(s)$, by Rule 2''.

Likewise, $\zeta(s)^k$ generates the number of ordered factorizations of n into k factors. If the factor 1 is regarded as inadmissible, then $(\zeta(s) - 1)^k$ generates the number of ordered factorizations of n in which there are k factors, all ≥ 2.

One can go on and study further examples of interesting number-theoretic sequences that are generated by relatives of the Riemann zeta function, but there is a somewhat breathtaking generalization that takes in all of these at a single swoop, so let's prepare the groundwork for that.

Definition. *A number-theoretic function is a function whose domain is the set of positive integers. A number-theoretic function f is said to be multiplicative if it has the property that $f(mn) = f(m)f(n)$ for all pairs of relatively prime positive integers m and n.*

Since every positive integer n is uniquely, apart from order, a product of powers of distinct primes,

$$n = p_1^{a_1} p_2^{a_2} \cdots p_r^{a_r},$$

(2.6.4)

it follows that *a multiplicative number-theoretic function is completely determined by its values on all powers of primes*. Indeed,

$$f(n) = f(p_1^{a_1}) f(p_2^{a_2}) \cdots f(p_r^{a_r}).$$

(2.6.5)

For instance, suppose that I have a certain function f in mind. It is multiplicative and, further, for every prime p and positive integer m we have $f(p^m) = p^{2m}$. Well then, it must be that $f(n) = n^2$ for all n, because if n is as shown in (2.6.4), then

$$f(n) = f(\prod_i p_i^{a_i}) = \prod_i f(p_i^{a_i})$$

$$= \prod_i p_i^{2a_i} = \left\{\prod_i p_i^{a_i}\right\}^2$$

$$= n^2,$$

as claimed.

Another, less obvious, example of a multiplicative function is $d(n)$, the number of divisors of n. For instance,

$$6 = d(12) = d(3 \cdot 4) = d(3)d(4) = 2 \cdot 3 = 6.$$

To see that $d(n)$ is multiplicative in general, let m and n be relatively prime positive integers. Then every divisor d of mn is *uniquely* the product of a divisor d' of m and a divisor d'' of n. Indeed, we can take $d' = gcd(d, m)$ and $d'' = gcd(d, n)$. Therefore the number of divisors of mn is the product of the number of divisors of m and the number of divisors of n, which was to be shown.

It is quite easy, therefore, to dream up examples of multiplicative functions: let your function f do anything it likes on the powers of primes, then declare it to be multiplicative, and walk away.

Multiplicative number-theoretic functions satisfy an amazing identity, which we will state, then prove, and then use.

Theorem 2.6.1. *Let f be a multiplicative number-theoretic function. Then we have the formal identity*

$$\sum_{n=1}^{\infty} \frac{f(n)}{n^s} = \prod_p \left\{1 + f(p)p^{-s} + f(p^2)p^{-2s} + f(p^3)p^{-3s} + \cdots\right\} \quad (2.6.6)$$

in which the product on the right extends over all prime numbers p.

Proof. Imagine, if you will, multiplying out the product that appears on the right side of (2.6.6). Each factor in that product is an infinite series. The product looks like this, when spread out in detail:

$$(1 + f(2)2^{-s} + f(2^2)2^{-2s} + f(2^3)2^{-3s} + \cdots) \times$$
$$(1 + f(3)3^{-s} + f(3^2)3^{-2s} + f(3^3)3^{-3s} + \cdots) \times$$
$$(1 + f(5)5^{-s} + f(5^2)5^{-2s} + f(5^3)5^{-3s} + \cdots) \times$$
$$(1 + f(7)7^{-s} + f(7^2)7^{-2s} + f(7^3)7^{-3s} + \cdots) \times \cdots$$

$$(2.6.7)$$

To multiply out a bunch of formal infinite series like this, we reach into the first parenthesis and pull out one term, for instance $f(2^3)2^{-3s}$. Then we reach into the second parenthesis, pull out one term, say $f(3)3^{-s}$ and multiply it by the one we got earlier. This gives us an accumulated product (so far) of

$$f(2^3)f(3)2^{-3s}3^{-s} = \frac{f(2^3)f(3)}{(24)^s}. \qquad (2.6.8)$$

Suppose, just as an example, that in all of the following parentheses we exercise our choice of one term by pulling out the term '1.' Then, as a result of having made all of those choices, one out of each parenthesis, the contribution to the answer would be the single term shown in (2.6.8) above.

Now here's the interesting part. *That* particular set of choices has produced a term that involves $(24)^{-s}$. What other sequence of choices of a single term out of each parenthesis would *also* lead to a net contribution that involves $(24)^{-s}$? The answer: *no other set of choices can do that.*

Indeed, if from the first parenthesis we choose any term other than $f(2^3)2^{-3s}$, then no matter what terms we pull out of all following parentheses, there is no way we will ever find the power of 2, namely 2^{-3s}, that occurs in $(24)^{-s}$. We need three 2's, and no other parenthesis has any 2's at all to offer, so we'd better take them when we have the chance.

Similarly, we need a factor of 3^{-s} in order to complete the formation of the term $(24)^{-s}$. There are no 3's available in any parenthesis other than the second one, and there, to get the right number of 3's, namely one, we had better take the term $f(3)3^{-s}$ that we actually chose.

Thus, the coefficient of $(24)^{-s}$ on the right side of (2.6.6) is just what we found, namely $f(2^3)f(3)$. Now since f is multiplicative, that's the same as $f(24)$. Hence the coefficient of $(24)^{-s}$ is $f(24)$. But that is just what the left side of (2.6.6) claims.

Let's say that again, using 'n' instead of '24.' Let n be some fixed integer, and let (2.6.4) be its factorization into prime powers. In order to obtain a term that involves n^{-s}, i.e., that involves $\prod p_i^{-a_i s}$, we are forced to choose the '1' term in every parenthesis on the right side of (2.6.7), except for those parentheses that involve the primes p_i that actually occur in n. Inside a parenthesis that belongs to p_i, we must choose the one and only term in which p_i is raised to the power with which it actually occurs in n, else we won't have a chance of getting n^{-s}. Thus we are forced to choose the term $f(p_i^{a_i})p_i^{-a_i s}$ out of the parenthesis that belongs to p_i. That means that the coefficient of n^{-s} in the end will be

$$\prod_i f(p_i^{a_i}) = f(n),$$

by (2.6.5). ∎

Let's look again at (2.6.6). One thing that is very apparent is that a multiplicative function is completely determined by its values on all prime powers. Indeed, on the right side of (2.6.6) we see only the values of f at prime powers, but on the left, all values appear.

Try an example of the theorem. Take the multiplicative function $f(n) = 1$ (all n). Then (2.6.6) says that

$$\zeta(s) = \prod_p \left\{ 1 + p^{-s} + p^{-2s} + \cdots \right\}$$

$$= \prod_p \left\{ \frac{1}{1 - p^{-s}} \right\} \tag{2.6.9}$$

$$= \frac{1}{\prod_p (1 - p^{-s})},$$

which is a fundamental factorization of the zeta function.

For another example, take the multiplicative function $\mu(n)$ whose values on prime powers are

$$\mu(p^a) = \begin{cases} +1, & \text{if } a = 0; \\ -1, & \text{if } a = 1; \\ 0, & \text{if } a \geq 2. \end{cases}$$

With this function substituted for f in (2.6.6), one sees that the once formidable series in the braces now has only two terms, and (2.6.6) reads

$$\sum_{n \geq 1} \frac{\mu(n)}{n^s} = \prod_p \{1 - p^{-s}\}. \tag{2.6.10}$$

An important fact emerges by comparison of (2.6.9) with (2.6.10): the series $\zeta(s)$ and the series on the left side of (2.6.10) are reciprocals of each other. Hence,

$$\frac{1}{\zeta(s)} = \sum_{n \geq 1} \frac{\mu(n)}{n^s},$$

or, what amounts to the same thing, $1/\zeta(s) \overset{Dir}{\longleftrightarrow} \{\mu(n)\}_1^\infty$.

The function $\mu(n)$ is the *Möbius function*, and it plays a central role in the analytic theory of numbers, because of the fact that it is generated by the reciprocal of the Riemann zeta function. For instance, watch this: Suppose we have two sequences $\{h_n\}_1^\infty$ and $\{b_n\}_1^\infty$, and suppose that these two sequences are connected by the following equations—

$$a_n = \sum_{d \backslash n} b_d \qquad (n \geq 1). \tag{2.6.11}$$

The question is, how can we invert these equations, and solve for the b's in terms of the a's?

Nothing to it. Let the Dsgf's of the two sequences be $A(s)$ and $B(s)$. Then, if we take a step into Generatingfunctionland, we see that (2.6.11) means

$$A(s) = B(s)\zeta(s)$$

by Rule 1″. Hence $B(s) = A(s)/\zeta(s)$, and then from Rule 1″ again,

$$b_n = \sum_{d\backslash n} \mu\left(\frac{n}{d}\right) a_d \qquad (n = 1, 2, 3, \ldots) \qquad (2.6.12)$$

This is the celebrated Möbius Inversion Formula. The reciprocal relationships (2.6.11) and (2.6.12) of the sequences mirror the reciprocal relationships of their Dsgf's $\zeta(s)$ and $1/\zeta(s)$.

Example 1. Primitive bit strings.

How many strings of n 0's and 1's are *primitive*, in the sense that such a string is *not* expressible as a concatenation of several identical smaller strings?

For instance, 100100100 is not primitive, but 1101 is.

There are a total of 2^n strings of length n. Suppose $f(n)$ of these are primitive. Every string of length n is *uniquely* expressible as a concatenation of some number, n/d, of identical *primitive* strings of length d, where d is a divisor of n.

Thus we have

$$2^n = \sum_{d\backslash n} f(d) \qquad (n = 1, 2, \ldots)$$

By (2.6.12) we have

$$f(n) = \sum_{d\backslash n} \mu(\frac{n}{d}) 2^d \qquad (n = 1, 2, \ldots) \qquad (2.6.13)$$

for the required number.

Example 2. Cyclotomic polynomials

Among the n roots of the equation $x^n = 1$, the *primitive* nth roots of unity are those that are not also mth roots of unity for some $m < n$. Thus the 4th roots of unity are $\pm 1, \pm i$, but ± 1 are roots of $x^2 = 1$, so they aren't primitive 4th roots.

In general, the nth roots of unity are $\{e^{2\pi i r/n}\}_{r=0}^{n-1}$, and the primitive ones are

$$\{e^{2\pi i r/n}\} \qquad (0 \le r \le n - 1; \gcd(r, n) = 1).$$

So for each n there are exactly $\phi(n)$ primitive nth roots of unity.

The equation whose roots are *all* n of the n roots of unity is obviously the equation $x^n - 1 = 0$. The question is this: what is the polynomial

$\Phi_n(x)$ of degree $\phi(n)$ whose roots are exactly the set of *primitive* nth roots of unity? In other words, what can be said about the polynomial

$$\Phi_n(x) = \prod_{\substack{0 \le r \le n-1 \\ \gcd(r,n)=1}} (x - e^{2\pi i r/n}) \qquad (n = 1, 2, 3, \ldots)?$$

The polynomials $\Phi_n(x)$ are called the *cyclotomic* ("circle-cutting") polynomials.

The important fact for answering this question is that

$$\prod_{d\backslash n} \Phi_d(x) = 1 - x^n \qquad (n = 1, 2, 3, \ldots). \qquad (2.6.14)$$

Indeed, the right side of the equation is the product of all possible factors $(\omega - x)$ where ω is an nth root of unity, primitive or not. But every nth root of unity is a *primitive* dth rot of unity for exactly one $d \le n$, and that d is a divisor of n.

In detail, if we have some nth root $\omega = e^{2\pi i r/n}$, then let $g = \gcd(r, n)$, $d = n/g$, and $r' = r/g$. Since $\omega = e^{2\pi i r'/d}$ we see that ω is a primitive dth root of unity and that $d\backslash n$. Thus every linear factor on the right side of (2.6.14) occurs in one and only one of the cyclotomic polynomials on the left side of (2.6.14), which proves the assertion.

From (2.6.14) we will obtain a fairly explicit formula for the $\Phi_n(x)$, by inverting the equation to solve for the Φ's. The form of the equation reminds us of the setup (2.6.11) for the Möbius inversion formula, but we have a product over divisors instead of a sum over divisors. A small dose of logarithms will convert products to sums, however, so we take the logarithm of both sides of (2.6.14), to get

$$\sum_{d\backslash n} \log \Phi_d(x) = \log (1 - x^n) \qquad (n = 1, 2, 3, \ldots). \qquad (2.6.14)$$

This is now precisely in the form (2.6.11), so we can use (2.6.12) to invert it, the result being

$$\log \Phi_n(x) = \sum_{d\backslash n} \mu(\frac{n}{d}) \log (1 - x^d) \qquad (n = 1, 2, 3, \ldots).$$

Finally, we exponentiate both sides to obtain our "fairly explicit formula,"

$$\Phi_n(x) = \prod_{d\backslash n} (1 - x^d)^{\mu(n/d)} \qquad (n = 1, 2, 3, \ldots). \qquad (2.6.15)$$

This is a good time to remember that the values of the Möbius function μ can only be ± 1 or 0. So the exponents on the right side of (2.6.15) tell us whether to omit a certain factor, which we do if $\mu = 0$, to put it in the numerator (if $\mu = 1$), or to put it in the denominator (if $\mu = -1$). For instance, $\Phi_{12}(x)$ is

$$(1-x)^{\mu(12)}(1-x^2)^{\mu(6)}(1-x^3)^{\mu(4)}(1-x^4)^{\mu(3)}(1-x^6)^{\mu(2)}(1-x^{12})^{\mu(1)}$$
$$= (1-x)^0(1-x^2)^1(1-x^3)^0(1-x^4)^{-1}(1-x^6)^{-1}(1-x^{12})^1$$
$$= \frac{(1-x^2)(1-x^{12})}{(1-x^4)(1-x^6)}$$
$$= 1-x^2+x^4,$$

which didn't look much like a polynomial at all until the very last step!

An important and beautiful fact about these polynomials is that the equation $\Phi_n(z) = 0$ can always be solved by radicals. That is, the solutions can always be obtained by a finite number of root extractions and rational operations. This is certainly not the case for general polynomial equations. As an example of this property we note the splendid fact that

$$\cos\frac{2\pi}{17} = \frac{1}{16}\Big\{-1+\sqrt{17}+\sqrt{2(17-\sqrt{17})}$$
$$+ 2\sqrt{17+3\sqrt{17}-\sqrt{2(17-\sqrt{17})}-2\sqrt{2(17+\sqrt{17})}}\,\Big\}.$$

The proof is fairly difficult, and can be found in Rademacher [Ra].

Some applications of cyclotomic polynomials will appear in section 4.9.

Exercises

1. Calculate the first three coefficients of the reciprocals of the power series of the functions:

    (a) $\cos x$

    (b) $(1 + x)^m$

    (c) $1 + t^2 + t^3 + t^5 + t^7 + t^{11} + \cdots$

2. Calculate the first three coefficients of the inverses of the power series for the functions:

    (a) $\sin x$

    (b) $\tan x$

    (c) $x + x^2 \sqrt{1 + x}$

    (d) $x + x^3$

    (e) $\log(1 - x)$

3. Let f be a formal power series such that $f'' + f = 0$. Give a careful proof that $f = A \sin x + B \cos x$.

4. Find simple closed formulas for the opsgf's of the following sequences:

    (a) $\{n + 7\}_0^\infty$

    (b) $\{1\}_4^\infty$

    (c) $\{1, 0, 1, 0, 1, 0, 1, 0, \ldots\}$

    (d) $\{1/(n + 1)\}_2^\infty$

    (e) $\{1/(n + 5)!\}_0^\infty$

    (f) $F_1, 2F_2, 3F_3, 4F_4, \ldots$ (the F's are the Fibonacci numbers)

    (g) $\{(n^2 + n + 1)/n!\}_1^\infty$

5. Use generating functions to prove that $\sum_k \binom{n}{k} = 2^n$.

6. Given positive integers n, k; define $f(n, k)$ as follows: for each way of writing n as an ordered sum of exactly k nonnegative integers, let S be the product of those k integers. Then $f(n, k)$ is the sum of all of the S's that are obtained in this way. Find the opsgf of f and an explicit, simple formula for it.

7. Let $f(n, k, h)$ be the number of ordered representations of n as a sum of exactly k integers, each of which is $\geq h$. Find $\sum_n f(n, k, h)x^n$.

8. Find the limit superior of each of the following sequences. In each case give a careful proof that your answer is correct.

    (a) $1, 0, 1, 0, 1, 0, \ldots$

    (b) $\{(-1)^n\}_0^\infty$

(c) $\{\cos(n\pi/k)\}_{n\geq 0}$ ($k \neq 0$ is a fixed integer)

(d) $\{1 + ((-1)^n/n)\}_{n\geq 1}$

(e) $\{n^{1/n}\}_{n\geq 1}$

9. Prove that if a sequence has a limit then its limit superior is equal to that limit.

10. Prove that a sequence cannot have two distinct limits superior.

11. Find the radius of convergence of each of the following power series:

(a) $\sum_{n\geq 1} x^n/(n^2)$

(b) $1 + x^3 + x^6 + x^9 + x^{12} + \cdots$

(c) $1 + 5x^2 + 25x^4 + 125x^6 + \cdots$

(d) $1 + 2!x^2 + 4!x^4 + 6!x^6 + \cdots$

(e) $\sum_{n\geq 0} x^{n!}$

12. Finish the proof of theorem 2.4.1 in the cases where $R = 0$ and $R = \infty$.

13. Show that if $\{f(n)\}_1^\infty$ is a multiplicative function, then so is

$$g(n) = \sum_{d\backslash n} f(d) \qquad (n = 1, 2, \ldots)$$

14. Euler's function $\phi(n)$ is the number of integers $1 \leq m \leq n$ such that m is relatively prime to n. Show by a direct counting argument that

$$\sum_{d\backslash n} \phi(d) = n \qquad (n = 1, 2, \ldots).$$

15. Show that each of the following functions is multiplicative. In each case find the value of the function when n is a prime power, and thereby find a formula for its value on any integer n.

(a) Euler's function $\phi(n)$ (use the results of problems 13, 14 above).

(b) $\sigma(n)$, which is the sum of the divisors of n.

(c) The function $|\mu(n)|$, which is 1 if n is not divisible by a square and 0 otherwise.

16. For each of the functions defined in problem 15 above, find its Dirichlet series generating function by using Theorem 2.6.1. First substitute into (2.6.6) the values of the function at prime powers. Then try to sum the power series that occurs, in closed form. Finally, by comparing the product that results with (2.6.9), try to express your answer simply in terms of the Riemann zeta function. In each case the Dsgf can be simply expressed in terms of $\zeta(s)$, or small variations thereof.

17. Find the Dsgf of each of the following sequences:

(a) $\{n\}_1^\infty$

(b) $\{n^\alpha\}_1^\infty$

(c) $\{\log n\}_1^\infty$

(d) $\{\sum_{d\backslash n} d^q\}_{n=1}^\infty$

18. For each of the following identities: first check that the identity is sometimes correct by calculating both sides of the alleged equation when $n = 1, 2, 3, 4, 5, 6, 7, 8$; next find the Dsgf's of the sequences on both sides of the claimed identity, observe that they are the same, and thereby prove the identity. Use the results of exercise 16 above.

(a) $\sum_{d\backslash n} \phi(d) = n$ $(n \ge 1)$

(b) $\sum_{d\backslash n} \mu(d) = 0$ if $n \ge 2$ and $=1$ if $n = 1$

(c) $\sum_{\delta\backslash n} \mu(\delta)d(n/\delta) = 1$ $(n \ge 1)$

19. If $\{f(k)\}$ is the sequence in example 7 of section 2.2, show that $f(k) = F_{2k-1}$ for $k \ge 1$, where the $\{F_k\}$ are the Fibonacci numbers.

20. Prove the binomial theorem

$$(x + y)^n = \sum_k \binom{n}{k} x^k y^{n-k}$$

by comparing the coefficient of $t^n/n!$ on both sides of the equation $e^{t(x+y)} = e^{tx}e^{ty}$. Prove the multinomial theorem

$$(x_1 + \cdots + x_h)^n = \sum_{r_1 + \cdots + r_k = n} \frac{n!}{r_1! \cdots r_k!} x_1^{r_1} \cdots x_h^{r_k}$$

by a similar device.

21.

(a) Let T be a fixed set of nonnegative integers. Let $f(n, k, T)$ be the number of ordered representations of n as a sum of k integers chosen from T. Find $\sum_n f(n, k, T)x^n$.

(b) Let $g(n, k, T)$ be the number of ordered representations of n as a sum of k *distinct* integers chosen from T. Find $\sum_n g(n, k, T)x^n$.

(c) Finally, let S, T be two fixed sets of nonnegative integers. Let $f(n, k, S, T)$ be the number of ordered representations of n as a sum of k integers chosen from T, each being chosen with a multiplicity that belongs to S. Find $\sum_n f(n, k, S, T)x^n$.

22. Let $f(n)$ be the excess of the number of ordered representations of n as the sum of an even number of positive integers over those as a sum of an odd number of them. Find $f(n)$ by finding $\sum_n f(n)x^n$ and reading off its coefficients.

23. Let $\{B_n\}$ be the sequence of Bernoulli numbers defined by (2.5.8), and let m be a positive integer. By considering the generating function

$$\frac{x(e^{mx} - 1)}{e^x - 1}$$

in two ways, find an evaluation of the sum of the rth powers of the first N positive integers as a polynomial of degree $r + 1$ in N, whose coefficients are given quite explicitly in terms of the Bernoulli numbers.

24.

(a) Make a table of values of the classical Möbius function $\mu(n)$ for $n = 1, 2, \ldots, 30$.

(b) Make a table of the values of the function $f(n)$ of (2.6.13) for $n = 1, 2, \ldots, 12$.

(c) Make a list of the primitive strings of length 6, and verify your value of $f(6)$.

25. This problem is intended to show how generating functions occur in coding theory. An important question in coding theory is the following: for fixed integers n and d, what is the length $A(n, d)$ of the longest list of n-bit strings of 0's and 1's (*codewords*) such that two distinct codewords always differ in at least d bit positions?

(a) Assign to each of the 2^n codewords $(\epsilon_1, \ldots, \epsilon_n)$ a *color*, as follows: the color of ϵ is $\sum_j j\epsilon_j$ modulo $2n$. Show that if two codewords differ in just 1 or 2 coordinates, then they are assigned distinct colors in this scheme.

(b) From part (a), show that $A(n, 3) \geq 2^n/(2n)$.

(c) Let a_j be the number of codewords for which $\sum_r r\epsilon_r = j$, for each j. Find the opsgf $f(z) = \sum_j a_j z^j$ explicitly as a product.

(d) If β_r is the number of codewords of color r, then express β_r in terms of the a_j's above. Then use the roots of unity method of section 2.4 to find that

$$\beta_r = \frac{1}{2n} \sum_{j=1}^{n''} 2^{2^a (j,n'')} e^{-\frac{2\pi i r j}{n''}}$$

for each $r = 0, 1, \ldots, 2n-1$. Here a and n'' are defined by $n = 2^a n''$ where n'' is odd, and (b, c) denotes the g.c.d. of b and c.

(e) Deduce that β_0 is the largest of the β_r's, and therefore find the stronger bound

$$A(n, 3) \geq \frac{1}{2n} \sum_{j=1}^{n''} 2^{2^a (j,n'')} \geq \frac{2^n}{2n}.$$

(f) Use Parseval's identity and the result of part (d) to find the variance of the occupancy numbers $\beta_0, \ldots, \beta_{2n-1}$. Make an estimate that shows that the variance is in some sense very small, so that this coloring scheme is shown to distribute codewords into color classes very uniformly.

26. Derive (2.5.7) from (2.5.6). That is, show that

$$\binom{-n}{k} = (-1)^k \binom{n+k-1}{k}.$$

27. Let $D(n)$ be the number of derangements of n letters, discussed in Example 4.

(a) Find, in simple explicit form, the egf of $\{D(n)\}_0^\infty$.

(b) Prove, by any method, that

$$D(n+1) = (n+1)D(n) + (-1)^{n+1} \qquad (n \geq 0; D(0) = 1)$$

(c) Prove, by any method, that

$$D(n+1) = n(D(n) + D(n-1)) \qquad (n \geq 1; D(0) = 1; D(1) = 0).$$

(c) Show that the number of permutations of n letters that have exactly 1 fixed point differs from the number with no fixed points by ± 1.

(d) Let $D_k(n)$ be the number of permutations of n letters that have exactly k fixed points. Show that

$$\sum_{k,n \geq 0} D_k(n) \frac{x^n y^k}{n!} = \frac{e^{-x(1-y)}}{1-x}.$$

28. Prove the following variation of the Möbius inversion formula. Let $\{a_n(x)\}$ and $\{b_n(x)\}$ be two sequences of functions that are connected by the relation

$$a_n(x) = \sum_{d \backslash n} b_{\frac{n}{d}}(x^d) \qquad (n = 1, 2, 3, \ldots).$$

Then we have

$$b_n(x) = \sum_{d \backslash n} \mu(\frac{n}{d}) a_d(x^{n/d}) \qquad (n = 1, 2, 3, \ldots).$$

29.

(a) Make a table of the values of $\phi(n)$ for $1 \leq n \leq 25$.

(b) As far as your table goes, verify that ϕ is a multiplicative function, by actual computation. Then check by actual computation from your table, that the result stated in exercise 14 above is true when $n = 20$ and $n = 24$.

(c) Let $n = p^a$ where p is a prime number. What is $\phi(n)$?

(d) Use the results above to find a general formula for $\phi(n)$ in terms of the prime factorization $n = p_1^{a_1} p_2^{a_2} \cdots p_k^{a_k}$ of n. Use your result to calculate $\phi(2592)$.

(e) Find the Dirichlet series generating function of $\phi(n)$, using (2.6.6), and express it in terms of the Riemann zeta function.

(f) Apply the Möbius inversion formula to the result of exercise 18(a), and thereby "solve" 18(a) for $\phi(n)$, to get an explicit formula for $\phi(n)$ that involves a sum of various values of the Möbius function.

(g) Show that your answers to parts (d) and (f) of this problem are identical, even though they look different.

30. Find the Dirichlet series generating functions for the sequences

(a) $a_n = \sqrt{n}$

(b) $a_n = |\mu(n)|$, where μ is the Möbius function.

(c) A number-theoretic function $f(n)$ is *strongly multiplicative* if it is true that $f(mn) = f(m)f(n)$ for all pairs m, n of positive integers. Let $\lambda(n)$ be the strongly multiplicative function that takes the value -1 on every prime, and $\lambda(1) = 1$. Find its Dsgf, and then prove that

$$\sum_{d \backslash n} \lambda(d) = \begin{cases} 1 & \text{if } n \text{ is a square;} \\ 0 & \text{otherwise.} \end{cases}$$

31. A *Lambert series* is a series of the form

$$f(x) = \sum_{n \geq 1} a_n \frac{x^n}{1 - x^n},$$

and we then say that f is the Lambert series gf for the sequence $\{a_n\}$. (Lambert series are only rarely used because they're hard to analyze.)

(a) Suppose f is the Lambert series gf of a sequence $\{a_n\}_1^\infty$, and the same f is the opsgf of a sequence $\{b_n\}_1^\infty$. Find the b's in terms of the a's.

(b) Thus prove the amazing identity

$$\sum_{n \geq 1} \frac{\mu(n)x^n}{1 - x^n} = x,$$

where again μ is the Möbius function.

(c) Find the Lambert series generating function of Euler's ϕ function.

32. Let $a = \{a_n\}_{n \geq 0}$ be a given sequence. Let S be the operator that transforms a into its sequence of partial sums: $(Sa)_n = a_0 + \cdots + a_n$, for $n \geq 0$.

(a) If f is the opsgf of a, what is the opsgf of Sa?

(b) If f is the opsgf of a and $r \geq 0$ what is the opsgf of $S^r a$?

(c) What is $S^r a$ if a is the sequence of all 1's?

(d) For a general sequence a, find an explicit formula, involving a single summation sign, for the nth member of the sequence $S^r a$.

(e) An unknown sequence a has the following property: if, beginning with a we iterate r times the operation S, of replacing the sequence by its sequence of partial sums, we obtain the sequence $\{1, 0, 0, \ldots\}$. Find a.

33.

(a) Write out the first twelve cyclotomic polynomials.

(b) If $n = p^a$ is a prime power, what is $\Phi_n(x)$?

(c) Show that for $n \geq 1$,

$$\Phi_n(1) = \begin{cases} 1, & \text{if } n > 1 \text{ is not a prime power;} \\ p, & \text{if } n = p^k \text{ is a prime power;} \\ 0, & \text{if } n = 1. \end{cases}$$

34. Consider the following sequence of polynomials.

$$\psi_n(x) = \sum_{\substack{1 \leq m \leq n \\ \gcd(m,n)=1}} x^m \qquad (n = 1, 2, \ldots).$$

Thus $\psi_1 = x$, $\psi_2 = x$, $\psi_3 = x + x^2$, $\psi_4 = x + x^3$, etc.

(a) Show that

$$\sum_{d \backslash n} \psi_{\frac{n}{d}}(x^d) = \frac{x(1 - x^n)}{1 - x} \qquad (n = 1, 2, \ldots).$$

(b) Use the result of exercise 28 to show that

$$\psi_n(x) = (1 - x^n) \sum_{d \backslash n} \mu(d) \frac{x^d}{1 - x^d} \qquad (n = 1, 2, \ldots).$$

(c) Show that at every primitive nth root of unity ω we have $\psi_n(\omega) = \mu(n)$, and therefore the polynomial $\psi_n(x) - \mu(n)$ is divisible by the nth cyclotomic polynomial $\Phi_n(x)$.

Chapter 3
Cards, Decks, and Hands: The Exponential Formula

3.1 Introduction

In this chapter we will discuss a particularly rich vein of applications of the theory of generating functions to counting problems. The exponential formula, which is our main goal here, is a cornerstone of the art of counting. It deals with the question of counting structures that are built out of connected pieces. The structures themselves need not be connected, but their pieces always are. The question is, if we know how many pieces of each size there are, how many structures of each size can we build out of those pieces?

We begin with a little example. There is 1 connected labeled graph that has 1 vertex, there is 1 connected labeled graph that has 2 vertices, and there are 4 connected labeled graphs that have 3 vertices. These creatures are all shown in Fig. 3.1 below.

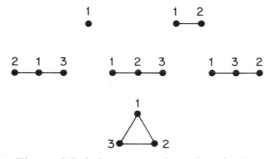

Fig. 3.1: The six labeled, connected graphs of ≤ 3 vertices.

Now think of graphs that have exactly 3 labeled vertices but are *not* necessarily connected. There are 8 of them, as shown in Fig. 3.2.

The question is, how can we develop a theory that will show us the connection between the number 8, of *all* graphs of ≤ 3 vertices, and the numbers 1, 1, 4 of *connected* labeled graphs of 1, 2, and 3 vertices? After all, such a theory should exist, because the connected graphs are the building blocks out of which all graphs are constructed. How, exactly, are those building blocks used?

Suppose we want to construct a graph G of n vertices and k connected components. We can first choose which k connected graphs to use for the connected components, subject only to the condition that the sum of their numbers of vertices must be n.

Second, after deciding which connected graphs to use, we need to relabel all of their vertices. That is because the connected graphs that we

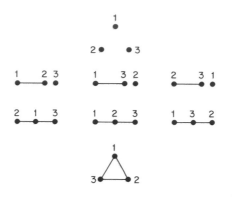

Fig. 3.2: The eight not-necessarily-connected labeled graphs of 3 vertices.

use, as in Fig. 3.1, each have their own private sets of vertex labels. A connected graph of 5 vertices will have labels 1, 2, 3, 4, 5 on its vertices, etc. However, the final assembled graph G, that we are manufacturing out of those connected pieces, will use each vertex label $1, 2, 3, \ldots, n$ exactly once, as in Fig. 3.2. Note, for instance, that the connected graph of 1 vertex appears three times in the first graph of Fig. 3.2 with 3 different vertex labels.

So our counting theory will have to take into account the choices of the connected graphs that are used as building blocks, as well as the number of ways to relabel the vertices of those connected graphs to obtain the final product.

Now we're going to raise the ante. Instead of going ahead and answering these counting questions in the case of graphs, it turns out to be better to be a bit more general right from the start, because a lot of nice applications don't quite fall under the heading of graphs. So we are going to develop the theory in a context of 'playing cards' and 'hands,' instead of 'connected graphs' and 'all graphs.'

Next you will see a number of definitions of the basic terminology. After all of those definitions, a few examples will no doubt be welcome, and will be immediately forthcoming. Then we will get on with the development of the theory and its numerous applications.

3.2 Definitions and a question

We suppose that there is given an abstract set P of 'pictures.'

Definition. A *card* $\mathcal{C}(S, p)$ is a pair consisting of a finite set S (the 'label set') of positive integers, and a picture $p \in P$. The *weight* of $\mathcal{C}$ is $n = |S|$. A card of weight n is called *standard* if its label set is $[n]$.*

---

* Recall that $[n]$ is the set $\{1, 2, \ldots, n\}$.

Definition. A *hand H* is a set of cards whose label sets form a partition of $[n]$, for some n.

This means that if n denotes the sum of the weights of the cards in the hand, then the label sets of the cards in H are pairwise disjoint, nonempty, and their union is $[n]$.

Definition. The *weight of a hand* is the sum of the weights of the cards in the hand.

Definition. A *relabeling* of a card $\mathcal{C}(S, p)$ with a set S' is defined if $|S| = |S'|$, and it is the card $\mathcal{C}(S', p)$. If $S' = [|S|]$ then we have the *standard* relabeling of the card.

Definition. A *deck* $\mathcal{D}$ is a finite set of standard cards whose weights are all the same and whose pictures are all different. The *weight of the deck* is the common weight of all of the cards in the deck.

Definition. An *exponential family* $\mathcal{F}$ is a collection of decks $\mathcal{D}_1, \mathcal{D}_2, \ldots$ where for each $n = 1, 2, \ldots$, the deck $\mathcal{D}_n$ is of weight n.

If $\mathcal{F}$ is an exponential family, we will write d_n for the number of cards in deck $\mathcal{D}_n$, and we will call $\mathcal{D}(x)$, the egf of the sequence $\{d_n\}_1^{\infty}$, the *deck enumerator* of the family.

Question: *Given an exponential family $\mathcal{F}$. For each $n \geq 0$ and $k \geq 1$, let $h(n, k)$ denote the number of hands H of weight n that consist of k cards, and are such that each card in the hand is a relabeling of some card in some deck in $\mathcal{F}$. Repetitions are allowed. That is, we are permitted to take several copies of the same card from one deck, and to relabel those copies with different label sets.*

How can we express $h(n, k)$ in terms of $d_1, d_2, d_3, \ldots$, where d_i is the number of different cards in deck $\mathcal{D}_i$ $(i \geq 1)$?

If $h(n, k)$ is the number of hands H of weight n that have exactly k cards, then we introduce the 2-variable generating function

$$\mathcal{H}(x, y) = \sum_{n,k \geq 0} h(n, k) \frac{x^n}{n!} y^k. \qquad (3.2.1)$$

This is a generator of mixed type; it is an opsgf with respect to the y variable and an *egf* with respect to x. We will call it the 2-variable *hand enumerator* of the family.

If $h(n) = \sum_k h(n, k)$ is the number of hands of weight n without regard to the number of cards in it, then we write $\mathcal{H}(x)$ for the egf of $\{h(n)\}$, instead of $\mathcal{H}(x, 1)$. It is the 1-variable hand enumerator of $\mathcal{F}$.

One way to answer the question raised above would be to exhibit a simple relationship between the generating functions $\mathcal{H}(x, y)$ and $\mathcal{D}(x)$, and that, of course, is exactly what we are about to do (see (3.4.4) below for a look at the answer).

3.3 Examples of exponential families

Before we get on with the business of answering the question that was raised in the previous section, here are a few examples of exponential families that have important roles in combinatorial theory.

Example 1.

The first exponential family that we will describe is the family of all vertex-labeled, undirected graphs. We will call this family $\mathcal{F}_1$.

A graph G is a set of vertices some pairs of which are designated as edges. A labeled graph is a graph that has a positive integer associated with each vertex. The integers ('labels') are all different. The graph has the *standard* labeling if the set of its vertex labels is $[n]$, where n is the number of vertices of G.

There are $\binom{n}{2}$ possible edges in graphs of n vertices, so there are $2^{\binom{n}{2}}$ labeled graphs of n vertices. For instance, there are 8 labeled graphs of 3 vertices, and these are shown in Fig. 3.2 (graphs are drawn by first drawing the n vertices and then, between each pair of vertices that is designated as an edge, drawing a line).

Some graphs are *connected* and some are *disconnected*. A graph is connected if, given any pair of vertices, we can walk from one to the other along edges in the drawing of the graph. Otherwise, the graph is disconnected. Of the 8 graphs of 3 vertices, shown in Fig. 3.2, 4 are connected, namely the last 4 that are pictured there.

Now let's describe our exponential family.

First, we describe a card $\mathcal{C}(S, p)$. There is a card corresponding to every *connected* labeled graph G. The set S is the set of vertex labels that is used in the graph.

Before we can describe the 'picture' on the card we need to say what a standard relabeling of a graph is. Let G be a graph of n vertices that are labeled with a set S of labels. Then relabel the vertices with $[n]$, *preserving the order of the labels*. That is, the vertex that had the smallest label in S will then get label 1, etc. Therefore the standard relabeling is uniquely defined.

Now, if G is a labeled, connected graph, the picture p on the card $\mathcal{C}(S, p)$ that corresponds to G is the standard relabeling of G. Hence, on a card $\mathcal{C}$ we see two things: a picture of a connected graph with standard labels, and another set of labels, of equal cardinality.

For instance, one card of weight 3 might be

$$(S, p) = \left(\{5, 9, 11\}, \quad \overset{1}{\bullet}\!\!-\!\!\overset{3}{\bullet}\!\!-\!\!\overset{2}{\bullet} \right)$$

which would correspond to the connected labeled graph

So cards correspond to connected graphs with not-necessarily-standard label sets.

What is a hand? A hand is a collection of cards whose label sets partition $[n]$, where n is the weight of the hand, which is to say, it is the total number of vertices in all of the connected graphs on all of the cards of the hand. But that is something very useful; a hand H corresponds to a not-necessarily-connected graph with standard labels! Its individual connected components may have nonstandard labels, but the graph itself uses exactly the labels $1, 2, \ldots, n$, where n is its number of vertices.

In summary then, the set of all vertex labeled graphs forms an exponential family. Each card is a labeled connected graph, each deck $\mathcal{D}_n$ is the set of all connected standard labeled graphs of n vertices, each hand is a standard (not-necessarily-connected) labeled graph. The number d_n of cards in the nth deck is the number of standard connected labeled graphs of n vertices, and the number $h(n, k)$ of hands of weight n with k cards is the number of standard labeled graphs of n vertices with k connected components. The question posed at the end of the last section in this case asks for the relationship between the numbers of *all* labeled graphs and all *connected* labeled graphs of all sizes. ∎

Example 2.

In this example we will find that the set of all permutations can be thought of as an exponential family.

First let's say what the cards are. On a card, the picture will show n points arranged in a circle, the points being labeled with the set $[n]$, in some order, and there will be arrowheads around the circle, all pointing clockwise, to tell us that the points are arranged in clockwise circular sequence.

So much for the 'picture' part of the card. Additionally, there is a set S of n positive integers on the card.

The reader will recognize that such a card corresponds to a *cyclic* permutation of the elements of S, i.e., a permutation of S that has a single cycle. For instance, the card whose picture is shown in Fig. 3.3

Fig. 3.3: A cyclic permutation is in the cards.

and whose set is $S = \{2, 4, 7, 9, 10\}$ represents the cyclic permutation

$$2 \longrightarrow 7 \longrightarrow 4 \longrightarrow 10 \longrightarrow 9 \longrightarrow 2$$

of the set S.

Now what is a *deck* of these cards? The cards in a deck are standard cards, and they consist of one sample of every distinct standard card of a given weight. In this case the nth deck $\mathcal{D}_n$ contains exactly $(n-1)!$ cards, one for each cyclic permutation of $[n]$.

So far we have accounted for the permutations with one cycle. They are the building blocks out of which all permutations are constructed, using hands of cards.

So what is a *hand*, in this example? A hand is a collection of cards whose label sets partition $[n]$, for some n. Thus we have a collection of cards in our hand in which the label sets are $S_1, \ldots, S_k$, say. These represent cyclic permutations of the S's, and the S's are pairwise disjoint with union $[n]$, where n is the number of points on all of the cards. A typical hand is shown below.

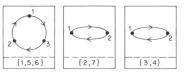

Since every permutation of n letters has a unique decomposition into cycles, we see that *hands of weight n correspond exactly to permutations of n letters.*

Hence the set of all permutations is an exponential family. We call it $\mathcal{F}_2$.

How many cards are in deck $\mathcal{D}_n$? There are $d_n = (n-1)!$ of them. The question raised at the end of the last section asks for the number $h(n,k)$ of hands of weight n and k cards. Such a hand represents a permutation of n letters that has k cycles. Hence in this case $h(n,k)$ is the number of permutations of n letters that have k cycles. When we have our general theorems in place, the ones that give the relationships between the d_n's and the $h(n,k)$'s, we'll learn a lot about permutations of various kinds with given numbers and sizes of cycles. Later, in chapter 5, we'll return to this subject and re-use these generating functions to get asymptotic information about permutations and their cycles.

3.4 The main counting theorems

In this section we will state and prove various forms of the exponential formula. The next section contains 10^9 applications of the method.

First, let two exponential families be given. We will say what it means to *merge* them. Roughly, it means to form a new family whose decks of each weight are the unions of the decks of those weights in the two given families. Some care is necessary, however, to insure that the two decks have all different cards, so we will now give a precise definition.

Let $\mathcal{F}'$ and $\mathcal{F}''$ be two exponential families whose picture sets P', P'' are disjoint. We form a third family $\mathcal{F}$, and write $\mathcal{F} = \mathcal{F}' \oplus \mathcal{F}''$, as follows:

fix $n \geq 1$. From $\mathcal{F}'$ we take all of the d_n' cards of deck $\mathcal{D}_n'$ and put them in a new pile. Then from $\mathcal{F}''$ we take all d_n'' of its cards from deck $\mathcal{D}_n''$ and add these d_n'' cards to the pile, which now contains $d_n = d_n' + d_n''$ different cards. Repeat this for each $n \geq 1$.

The Fundamental Lemma of Labeled Counting. *Let $\mathcal{F}'$, $\mathcal{F}''$ be two exponential families, and let $\mathcal{F} = \mathcal{F}' \oplus \mathcal{F}''$ be their merger. Further, let $\mathcal{H}'(x,y)$, $\mathcal{H}''(x,y)$, $\mathcal{H}(x,y)$ be the respective 2-variable hand enumerators of these families. Then*

$$\mathcal{H}(x,y) = \mathcal{H}'(x,y)\mathcal{H}''(x,y).$$

Proof. Consider a hand H in the merged family $\mathcal{F}$. Some of its cards came from $\mathcal{F}'$ and some came from $\mathcal{F}''$. The collection of cards that came from $\mathcal{F}'$ forms a sub-hand H' of weight, say, n', and having k' cards, that has been relabeled, in an order-preserving way, with a certain label set $S \subset [n]$. All hands H in the merged family are uniquely determined by a particular hand H' from $\mathcal{F}'$, the choice of new labels S with which that hand is to be relabeled, and the remaining subhand $\mathcal{H}''$ from $\mathcal{F}''$, which must be relabeled, again preserving the order of the labels, with $[n] - S$.

Consequently the number of hands in the merged family that have weight n and have exactly k cards is

$$h(n,k) = \sum_{n',k'} \binom{n}{n'} h'(n',k') h''(n-n',k-k')$$

$$- \left[\frac{x^n}{n!} y^k \right] \mathcal{H}'(x,y) \mathcal{H}''(x,y),$$

(3.4.1)

and we are finished. ■

The main idea is that the processes of merging families and of multiplying egf's correspond exactly. The fact that in equation (3.4.1) the n' variable in the sum carries a binomial coefficient along in its wake, while the k' does not, accounts for the mixed nature of the generating function that was chosen, with the 'x' variable being egf-like and the 'y' variable ops-like.

The Fundamental Lemma will allow us to build up the general relationship between deck and hand enumerators very easily, in a 'Sorcerer's Apprentice' fashion, beginning with a trickle and ending with a flood. We begin with a starkly simple exponential family that consists of exactly one nonempty deck that has just one card in it. The hand enumerator there will be obvious. Then we consider a family that has a number of cards in one deck, and no other decks. Finally we jump to the general situation, at each stage using the Fundamental Lemma, because we will be carrying out a merging operation.

Step 1: The trickle.

Fix a positive integer r. Let the rth deck, $\mathcal{D}_r$, contain exactly one card, and let all other decks be empty. The deck counts are $d_r = 1$ and all other $d_j = 0$. The deck enumerator is $\mathcal{D}(x) = x^r/r!$. A hand H consists of some number, say s, of copies of the one card that exists. The weight of H is rs. Therefore the number of hands of k cards and of weight n is $h(n,k) = 0$ unless $n = kr$. If $n = kr$, then how many hands of weight n are there? We can choose the labels for the first card in $\binom{n}{r}$ ways, for the second in $\binom{n-r}{r}$ ways, etc, for the kth card in $\binom{n-(k-1)r}{r} = 1$ way. Since the order of the labeled cards is immaterial, the number of hands is therefore

$$h(kr, k) = \frac{1}{k!} \frac{n!}{r!^k}.$$

The hand enumerator of this elementary family is therefore

$$\begin{aligned}
\mathcal{H}(x,y) &= \sum_{n,k} h(n,k) x^n y^k /n! \\
&= \sum_k \frac{x^{kr} y^k}{k! r!^k} \qquad\qquad (3.4.2) \\
&= \exp\left\{ \frac{yx^r}{r!} \right\}.
\end{aligned}$$

We won't have to do any more computation to get the general result; the Fundamental Lemma will do it for us.

Step 2: The flow

Fix positive integers r and d_r, and consider an exponential family $\mathcal{F}$ that has d_r cards in its rth deck $\mathcal{D}_r$, and has no other nonempty decks. We claim that the hand enumerator of this family is

$$\mathcal{H}(x,y) = \exp\left\{ \frac{yd_r x^r}{r!} \right\}. \qquad\qquad (3.4.3)$$

The proof is by induction on d_r. The claim is correct when $d_r = 1$, for that is (3.4.2). Suppose the claim is true for $d_r = 1, 2, \ldots, m-1$, and let the family $\mathcal{F}$ have m cards in its rth deck. Then $\mathcal{F}$ is the result of merging a family with $m-1$ cards in the rth deck and a family with 1 card in that deck. By the inductive hypothesis and the Fundamental Lemma, the hand enumerator is the product

$$\exp\left\{ y(m-1)x^r/r! \right\} \exp\left\{ yx^r/r! \right\} = \exp\left\{ ymx^r/r! \right\},$$

and the claim is proved.

Step 3: The flood.

We are now ready to prove the main counting theorem.

Theorem 3.4.1 (The exponential formula). *Let $\mathcal{F}$ be an exponential family whose deck and hand enumerators are $\mathcal{D}(x)$ and $\mathcal{H}(x, y)$, respectively. Then*

$$\mathcal{H}(x, y) = e^{y\mathcal{D}(x)}. \tag{3.4.4}$$

In detail, the number of hands of weight n and k cards is

$$h(n, k) = \left[\frac{x^n}{n!}\right]\left\{\frac{\mathcal{D}(x)^k}{k!}\right\}. \tag{3.4.5}$$

Proof. In (3.4.3) we have proved this result in the special case where there is only one nonempty deck. But a general exponential family with a full sequence of nonempty decks $\mathcal{D}_1, \mathcal{D}_2, \ldots$ is the merger of the special families $\mathcal{F}_r$ ($r = 1, 2, \ldots$), each of which has just a single nonempty deck $\mathcal{D}_r$. By the Fundamental Lemma, the hand enumerator of the general family is the product of the hand enumerators of the special families. But the generating function (3.4.4) claimed in the theorem is indeed the product of the enumerators (3.4.3) of the special families $\mathcal{F}_r$, and the proof is finished. ∎

By summing (3.4.5) over all k we obtain the following:

Corollary 3.4.1. *Let $\mathcal{F}$ be an exponential family, let $\mathcal{D}(x)$ be the egf of the sequence $\{d_n\}_1^\infty$ of sizes of the decks, and let $\mathcal{H}(x) \overset{egf}{\longleftrightarrow} \{h_n\}_0^\infty$, where h_n is the number of hands of weight n. Then*

$$\mathcal{H}(x) = e^{\mathcal{D}(x)}. \tag{3.4.6}$$

By summing (3.4.5) over just those k that lie in a given set T, we obtain

Corollary 3.4.2 (The exponential formula with numbers of cards restricted). *Let T be a set of positive integers, let $e_T(x) = \sum_{n\in T} x^n/n!$, and let $h_n(T)$ be the number of hands whose weight is n and whose number of cards belongs to the allowable set T. Then*

$$\{h_n(T)\}_0^\infty \overset{egf}{\longleftrightarrow} e_T(\mathcal{D}(x)). \tag{3.4.7}$$

The next several sections of this chapter will contain applications of the exponential formula.

3.5 Permutations and their cycles

We apply the theorems to the exponential family $\mathcal{F}_2$ of permutations, that was described in example 2 of section 3.3. There we observed that the deck $\mathcal{D}_n$ contains $d_n = (n-1)!$ cards. The exponential generating function of the sequence $\{(n-1)!\}_1^\infty$ is

$$\mathcal{D}(x) = \sum_{n\geq 1}(n-1)!\frac{x^n}{n!}$$

$$= \sum_{n\geq 1}\frac{x^n}{n}$$

$$= \log\frac{1}{1-x}.$$

Now from theorem 3.4.1 we have

$$\mathcal{H}(x,y) = \exp\left\{y\log\frac{1}{1-x}\right\}$$

$$= \frac{1}{(1-x)^y}. \tag{3.5.1}$$

In this exponential family, $h(n,k)$ is the number of permutations of n letters that have k cycles, and it is called the *Stirling* number of the first kind. We will use one of the standard notations, $\begin{bmatrix}n\\k\end{bmatrix}*$, for these numbers, and will reserve the $h(n,k)$ for the general situation.

Now,

$$\sum_k \begin{bmatrix}n\\k\end{bmatrix}y^k = \left[\frac{x^n}{n!}\right](1-x)^{-y}$$

$$= n!\binom{y+n-1}{n} \qquad \text{(by (2.5.7))} \tag{3.5.2}$$

$$= y(y+1)\cdots(y+n-1),$$

so the numbers of permutations of n letters with various numbers of cycles are the coefficients in the expansion of the 'rising factorial' function $y(y+1)\cdots(y+n-1)$.

The enumerator of hands of k cards is obviously

$$\frac{1}{k!}\left\{\log\frac{1}{1-x}\right\}^k \qquad (k=1,2,\ldots),$$

which tells us that the Stirling number is also given by

$$\begin{bmatrix}n\\k\end{bmatrix} = \left[\frac{x^n}{n!}\right]\frac{1}{k!}\left\{\log\frac{1}{1-x}\right\}^k. \tag{3.5.3}$$

---

* There are as many notations for $\begin{bmatrix}n\\k\end{bmatrix}$ as there are books on combinatorics. It is called $(-1)^k s(n,k)$ or $s_1(n,k)$, or $s(n,k)$, or $c(n,k)$, or several other things. Similarly the $\begin{Bmatrix}n\\k\end{Bmatrix}$ are called $s_2(n,k)$ or $S(n,k)$, etc.

One thing that we don't find is a simple little formula for these Stirling numbers. One can find formulas for them, but they're fairly unpleasant, involving double sums of summands with sign alternations, etc. But with the generating function apparatus we can do just about whatever we want to without such a formula. To calculate numerical values of the $\left[\begin{smallmatrix}n\\k\end{smallmatrix}\right]$, for instance, one can use the very simple recurrence relations that can be derived from these generating functions (see Exercise 8).

3.6 Set partitions

We introduce a new exponential family $\mathcal{F}_3$, as follows: first, for each $n \geq 1$, in the deck $\mathcal{D}_n$ there is just *one* card of weight n. On that card there is a picture of a smiling rabbit,* and there is the label set $[n]$.

What is a hand? There is a hand H corresponding to every partition of the set $[n]$. Indeed, given such a partition, take the sets in it and let them relabel the label sets on the cards in the hand. Then the cards are otherwise uniquely determined since there's only one card of each weight.

So in this exponential family the number of hands of weight n that have k cards is equal to the number of partitions of the set $[n]$ into k classes. But that is something we've met before, in example 6 of chapter 1, where we called those numbers $\left\{\begin{smallmatrix}n\\k\end{smallmatrix}\right\}$, the Stirling numbers of the second kind.

To apply the exponential formula we first compute the egf of the numbers d_n of cards in each deck. But these numbers are all 1, if $n \geq 1$, and are 0 else, so

$$\mathcal{D}(x) = \sum_n d_n \frac{x^n}{n!} = \sum_{n \geq 1} \frac{x^n}{n!} = e^x - 1.$$

Now by the exponential formula the enumerator of hands is

$$\mathcal{H}(x,y) = e^{y(e^x - 1)}, \tag{3.6.1}$$

and in particular

$$\left\{\begin{matrix}n\\k\end{matrix}\right\} = \left[\frac{x^n}{n!}\right]\left\{\frac{(e^x - 1)^k}{k!}\right\}. \tag{3.6.2}$$

Compare this result with the generating function (1.6.12) of the Bell numbers and find that we have here a refinement of that generating function. Not only does $e^{e^x - 1}$ generate the numbers of partitions of n-sets, but *each term* of the expansion

$$e^{e^x - 1} = \sum_{k \geq 0} \frac{(e^x - 1)^k}{k!}$$

has significance with respect to the numbers of classes in the partitions.

---

* Why not? Since there's only one card the picture is immaterial, so it might as well be cheerful.

3.7 A subclass of permutations

How many permutations σ of n letters have the property that σ has an even number of cycles and all of them are of odd lengths?

This problem takes place in an exponential family that is like the family $\mathcal{F}_2$ of permutations, except that it contains only the decks of odd weights, $\mathcal{D}_1, \mathcal{D}_3, \ldots$. The numbers $\{d_n\}_1^\infty$ that count the cards in the decks are now $1, 0, 2, 0, 24, 0, 720, \ldots$. The egf of the deck counts is

$$D(x) = \sum_{n \text{ odd}} (n-1)! \frac{x^n}{n!}$$

$$= \sum_{r \geq 0} \frac{x^{2r+1}}{2r+1}$$

$$= \log \sqrt{\frac{1+x}{1-x}}$$

by (2.5.2).

Since the number of cycles is required to be even, the allowable numbers of cards in a hand are the set T=the even numbers. By (3.4.7), the egf of the answer is

$$\cosh \left\{ \log \sqrt{\frac{1+x}{1-x}} \right\} = \frac{1}{\sqrt{1-x^2}}$$

$$= \sum_{m \geq 0} \binom{2m}{m} (x/2)^{2m}.$$

The number of permutations that meet the conditions of the problem is the coefficient of $x^n/n!$ here, namely

$$\binom{n}{\frac{n}{2}} \frac{n!}{2^n}.$$

That's one way to answer the question, but the answer can be restated in quite a striking form, like this-

Theorem 3.7.1. *Let a positive integer n be fixed. The probabilities of the following two events are equal:*
   (a) a permutation is chosen at random from among those of n letters, and it has an even number of cycles, all of whose lengths are odd
   (b) a coin is tossed n times and exactly $n/2$ heads occur.

3.8 Involutions, etc.

Fix positive integers m, n. How many permutations σ, of n letters, satisfy $\sigma^m = 1$, where '1' is the identity permutation?

To do this problem, we need the following:

Lemma. *For $\sigma^m = 1$ it is necessary and sufficient that all of the cycle lengths of σ be divisors of m.*

Proof. Consider a cycle C of σ, of length r. Let i be some letter that is in C. Then, by definition of a cycle, $\sigma^m(i)$ is the letter on C that we encounter by beginning at i and moving m steps around the cycle, namely the letter that is $m \bmod r$ steps around C from i. But $\sigma^m(i) = i$. Therefore $m \bmod r = 0$, i.e., r divides m. Therefore m is a multiple of the length of every cycle of C. The converse is clear, and the proof is finished. ∎

Now back to the problem. Consider the exponential family $\mathcal{F}_4$ in which the cards are the usual ones for cycles of permutations, but in which the only decks that occur are those whose weights are divisors of m. Then $d_r = (r-1)!$ if $r \backslash m$, and is 0 else. Hence

$$\mathcal{D}(x) = \sum_{r \geq 1} d_r x^r / r! = \sum_{d \backslash m} \frac{x^d}{d}. \qquad (3.8.1)$$

By the exponential formula (theorem 3.4.1) we have the following elegant result:

Theorem 3.8.1. *Fix $m > 0$. The numbers of permutations of n letters whose mth power is the identity permutation have the generating function*

$$\exp\left(\sum_{d \backslash m} (x^d / d)\right). \qquad (3.8.2)$$

∎

Let's try a special case of this theorem. Take $m = 2$. Then we are talking about permutations whose square is 1. These are called *involutions*. Involutions can have cycles of lengths 1 or 2 only, by the lemma above. If t_n is the number of involutions of n letters, then by (3.8.2) we have

$$\sum_{n \geq 0} \frac{t_n}{n!} x^n = e^{x + \frac{1}{2}x^2}. \qquad (3.8.3)$$

3.9 2-regular Graphs

How many undirected, labeled graphs are there on n vertices, in which every vertex is of degree 2 (such graphs are called *2-regular*)?

Such a graph is a disjoint union of undirected cycles, so we have an exponential family $\mathcal{F}_5$ in which the cards stand for undirected cycles, instead of directed ones, as in the case of permutations.

For fixed $n \leq 2$ there are no undirected cycles at all. For $n \geq 3$, the number d_n of cards in the nth deck is the number of undirected circular arrangements of n letters, and that number is $(n-1)!/2$. Therefore the generating function of the deck sizes is

$$
\begin{aligned}
\mathcal{D}(x) &= \sum_{n \geq 3} \frac{(n-1)!}{2\,n!} x^n \\
&= \frac{1}{2} \sum_{n \geq 3} x^n / n \\
&= \frac{1}{2} \left\{ \log \frac{1}{1-x} - x - \frac{x^2}{2} \right\}.
\end{aligned}
$$

By the exponential formula (3.4.4), the exponential generating function of the number $g(n)$ of undirected 2-regular labeled graphs is

$$
\begin{aligned}
\sum_{n \geq 0} g(n) \frac{x^n}{n!} &= \exp \left\{ \frac{1}{2} \log \frac{1}{1-x} - \frac{x}{2} - \frac{x^2}{4} \right\} \\
&= \frac{e^{-\frac{1}{2}x - \frac{1}{4}x^2}}{\sqrt{1-x}}.
\end{aligned}
\tag{3.9.1}
$$

This answer is a sparkling example of the ability of the generating function method to produce answers to difficult counting problems with minimal effort.

3.10 Counting connected graphs

How many labeled, connected graphs of n vertices are there?

Now we're back in the exponential family $\mathcal{F}_1$ of labeled graphs, but there are one or two little twists. The exponential formula can tell you the number of all gadgets of each size if you know the number of connected ones, or vice versa. This problem is 'vice versa.' The number of all labeled graphs of n vertices is $2^{\binom{n}{2}}$, so in the equation 'Hands $= e^{\text{Decks}}$' we know 'Hands' and we want to find 'Decks,' rather than the other way around.

There's one more twist. Let $\mathcal{D}(x)$ and $\mathcal{H}(x)$ be the egf's of the decks and the hands, respectively. Then

$$
\mathcal{H}(x) = \sum_{n \geq 0} \frac{2^{\binom{n}{2}}}{n!} x^n,
$$

and this series does not converge for any $x \neq 0$. So this is a formal power series generating function only, and we should not expect analytic functions at the end of the road.

Having said all of that, the machinery still works very nicely. We will now find a recurrence formula for the number of connected graphs by the '$xD \log$' method of section 1.6. It isn't any harder to find a general recurrence relation than for this special case, however, so let's do it in general.

Theorem 3.10.1. *The counting sequences $\{d_n\}$ and $\{h_n\}$, of decks and hands in an exponential family satisfy the recurrence*

$$nh_n = \sum_k \binom{n}{k} k d_k h_{n-k} \qquad (n \geq 1; h_0 = 1). \qquad (3.10.1)$$

Proof. Apply the '$xD \log$' method of section 1.6 to the exponential formula (3.4.4). ∎

It follows that *the numbers d_n of connected labeled graphs of n vertices satisfy the recurrence*

$$n2^{\binom{n}{2}} = \sum_k \binom{n}{k} k d_k 2^{\binom{n-k}{2}} \qquad (n \geq 1). \qquad (3.10.2)$$

From this formula we are able, for example, to compute the d_n's for small n. For $n = 1, \ldots, 6$ we find the values 1, 1, 4, 38, 728, 26704.

3.11 Counting labeled bipartite graphs

How many bipartite vertex-labeled graphs of n vertices are there?

The exponential formula can handle even this problem with just a little bit of coaxing. A *bipartite graph* G is a graph whose vertex set $V(G)$ can be partitioned into $V = A \cup B$ such that every edge of G is of the form (a, b), where $a \in A$ and $b \in B$. A bipartite graph of 10 vertices is shown in Fig. 3.4.

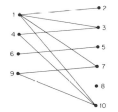

Fig. 3.4: A bipartite graph

Now, of the $2^{\binom{n}{2}}$ labeled graphs of n vertices, how many are bipartite?

Well, there's a little problem. The exponential formula can count the hands if you can count the decks, or it can count the decks if you can count the hands. But it can't do both, and in this problem it isn't immediately clear how many *connected* bipartite graphs there are *or* how many there are altogether.

A thought might be to choose the sets A, B of the partition $[n] = A \cup B$, and then count the bipartite graphs that have that partition. The latter is easy; since there are $|A||B|$ possible edges, there must be $2^{|A||B|}$ ways to exercise the freedom to draw or not to draw all of those edges.

The problem is that a fixed bipartite graph might get counted several times in the process. In other words, there may be several ways to exhibit a partition of the vertex set with all edges running between vertices in different classes. For instance, the graph G of Fig. 3.4 would turn up several times: once with $A = \{1, 4, 6, 9\}$, again with $A = \{2, 3, 5, 7, 8, 10\}$, again with $A = \{1, 4, 5, 9\}$, etc.

In general, a bipartite graph that has c connected components would be created 2^c times by the construction that we are considering, the reason being that for each connected component G_i of G we can choose which of the two sets in its vertex partition, A_i or B_i, will get put on the left hand side, in A, and which on the right hand side, in B.

To get around this conundrum we use slightly different playing cards. By a *2-colored* bipartite graph we mean a vertex-labeled bipartite graph G together with a coloring of the vertices of G in two colors ('Red,' 'Green'), such that whenever (v, w) is an edge of G, then v and w have different colors.

A *connected* bipartite graph, for instance, creates two 2-colored graphs. A bipartite graph with c connected components creates 2^c such 2-colored graphs.

In the exponential family $\mathcal{F}_6$ that we are making, there will be a *card* $\mathcal{C}$ corresponding to each 2-colored *connected* labeled bipartite graph. Imprinted on the card there will be, as always, S, the set of vertex labels that are used, and a picture of a 2-colored, connected bipartite graph of $|S|$ vertices with standard vertex labels.

What have we gained by coloring the cards? Just this: we now know how many hands of weight n there are. That number is

$$\gamma_n = \sum_k \binom{n}{k} 2^{k(n-k)}, \qquad (3.11.2)$$

because each and every hand arises exactly once from the following construction:

    (i) fix an integer k, $0 \le k \le n$.

    (ii) choose k of the elements of $[n]$ and color them 'Red.'

    (iii) color the remaining elements of $[n]$ 'Green.'

    (iv) decide independently for each vertex pair (ρ, γ), where ρ is Red and γ is Green, whether or not to make (ρ, γ) an edge.

It is obvious that (3.11.2) counts the possible outcomes of the construction.

So, even though we are in the wrong exponential family, because things are colored that we wish weren't, at least we know how many hands there are!

Next, let's use the exponential formula to find the egf for the decks, which correspond to *connected* 2-colored bipartite graphs. It tells us instantly that

$$\mathcal{D}(x) = \log\left\{\sum_{n\geq 0}\frac{\gamma_n}{n!}x^n\right\}, \tag{3.11.3}$$

where γ_n is defined by (3.11.2).

Now that we have the *connected* colored graphs counted, is it hard to count the *connected* uncolored graphs? Not at all, because there are just half as many uncolored and connected as there are colored and connected. So the egf of ordinary, uncolored connected bipartite graphs is $\mathcal{D}(x)/2$, where $\mathcal{D}(x)$ is given by (3.11.3).

But now we have achieved, in the *correct* exponential family, the objective that we had not reached before: we know how many cards there are in each deck. So we know one of the two items that the exponential formula relates, and therefore we can find the other one.

Since $\mathcal{D}(x)/2$ generates the deck counts, it must be that

$$\begin{aligned} e^{\mathcal{D}(x)/2} &= \exp\left\{\frac{1}{2}\log\left\{\sum_{n\geq 0}\frac{\gamma_n}{n!}x^n\right\}\right\} \\ &= \sqrt{\sum_{n\geq 0}\frac{\gamma_n}{n!}x^n} \end{aligned} \tag{3.11.4}$$

generates the hand counts, and we have:

Theorem 3.11.1. *Let $\beta(n)$ denote the number of vertex labeled bipartite graphs of n vertices. Then*

$$\sum_{n\geq 0}\frac{\beta(n)}{n!}x^n = \sqrt{\sum_{n\geq 0}\frac{\gamma_n}{n!}x^n}, \tag{3.11.5}$$

where the γ_n are given by (3.11.2).

So all of the complications about multiple counting were resolved by taking the square root of the generating function that we started with!

3.12 Counting labeled trees

A tree is a connected graph that has no cycles. How many (standard) labeled trees of n vertices are there?

In this example we will derive the answer to that question in the form of one of the most famous results in combinatorics, namely:

Theorem 3.12.1. *For each $n \geq 1$ there are exactly n^{n-2} labeled trees of n vertices.*

Although many proofs are known, the one by generating functions, which uses the exponential formula, is particularly enchanting, and here it is:

A *rooted* tree is a tree that has a distinguished vertex called the root. There are obviously n times as many labeled rooted trees of n vertices as there are trees, so we will be finished if we can count the rooted ones.

Let t_n be the number of rooted trees (with standard labels) of n vertices for $n \geq 1$. We define an exponential family $\mathcal{F}_7$ as follows. The cards correspond to rooted labeled trees. On a card $\mathcal{C}(S, p)$, p is a picture of a standard rooted tree of $|S|$ vertices, and S is a set of labels.

In $\mathcal{F}_7$, what is a hand? A hand H corresponds to a *rooted labeled forest*, which is a labeled graph each of whose connected components is a rooted tree. The exponential formula will tell us how many forests there are if we know how many trees there are, or vice versa. But this is one of those unsettling situations where we know neither. The solution? Press on, and keep the faith.

By the exponential formula,

$$\mathcal{H}(x) = e^{\mathcal{D}(x)}, \tag{3.12.1}$$

where $\mathcal{H}(x) \overset{egf}{\longleftrightarrow} \{f_n\}$, $\mathcal{D}(x) \overset{egf}{\longleftrightarrow} \{t_n\}$ and f_n is the number of rooted forests of n vertices. Now (3.12.1) is one equation in two unknown functions. To get another one we use a fact that was discovered by Pólya, namely that

$$t_{n+1} = (n+1)f_n \qquad (n \geq 0). \tag{3.12.2}$$

To prove (3.12.2), let F be a rooted labeled forest of n vertices. Introduce a new vertex v, and assign to it a label j, where $1 \leq j \leq n+1$. Relabel F with the set $1, 2, \ldots, j-1, j+1, \ldots, n+1$, preserving the order of the labels. Then draw edges between v and all of the roots of the components of F, and root the resulting tree at v. The result is a rooted labeled tree of $n+1$ vertices. As we vary the label j, we construct $n+1$ rooted trees corresponding to each rooted forest F. The construction is easily reversible, so every rooted tree of $n+1$ vertices occurs exactly once, which proves (3.12.2). ■

The sequence $f_n = t_{n+1}/(n+1)$ has the egf

$$\mathcal{H}(x) = \sum_{n \geq 0} \frac{f_n}{n!} x^n$$

$$= \sum_{n \geq 0} \frac{t_{n+1}}{(n+1)!} x^n$$

$$= \frac{1}{x} \mathcal{D}(x).$$

If we combine this with (3.12.1) we get

$$\mathcal{D}(x) = x e^{\mathcal{D}(x)}. \tag{3.12.3}$$

Now, in previous problems where there was an unknown generating function it has always happened that we obtained some sort of functional equation that had to be solved in order to find the function. We have seen situations where the equation was a differential equation, and others where it was a quadratic equation. In (3.12.3) we have a functional equation that is to be solved for $\mathcal{D}(x)$, which in fact determines $\mathcal{D}(x)$ uniquely, but which is not a differential equation or an algebraic equation, and whose solution isn't obvious at all.

There is a powerful tool for dealing with this kind of a functional equation, called the Lagrange Inversion Formula, which will be discussed in section 5.1. There we will finish the enumeration of trees as an illustration of the use of the Lagrange formula.

3.13 Exponential families and polynomials of 'binomial type.'

Associated with each exponential family there is a sequence of polynomials

$$\phi_n(y) = \sum_k h(n,k)y^k \qquad (n = 0, 1, 2, \ldots), \qquad (3.13.1)$$

where $h(n, k)$ is the number of hands of weight n and k cards. In view of the exponential formula (3.4.4) these polynomials satisfy the generating relation

$$e^{y\mathcal{D}(x)} = \sum_{n \geq 0} \frac{\phi_n(y)}{n!} x^n. \qquad (3.13.2)$$

Polynomial sequences that satisfy (3.13.2) have been called *polynomials of binomial type* by Rota and Mullin [RM]. The reason for the name is that since

$$e^{u\mathcal{D}(x)} \xrightarrow{egf} \{\phi_n(u)\}; \quad e^{v\mathcal{D}(x)} \xrightarrow{egf} \{\phi_n(v)\};$$

it follows that

$$\phi_n(u+v) = \sum_r \binom{n}{r} \phi_r(u)\phi_{n-r}(v) \qquad (n \geq 0),$$

which is reminiscent of the binomial theorem.

Although various authors have given combinatorial interpretations for such polynomial sequences, the very natural interpretation that appears above seems not to have been discussed. That interpretation is: *when the coefficients of polynomials $\{\phi_n(y)\}$ of binomial type are nonnegative, then there exists an exponential family $\mathcal{F}$ such that for each $n \geq 0$, $\phi_n(y)$ generates the hands of weight n, by numbers of cards.* Conversely, every exponential family has a family of polynomials of binomial type associated with it.

3.14 Unlabeled cards and hands

In the remainder of this chapter we will consider the same kinds of problems, except that there will be no label sets to worry about. This would seem to simplify things, and it does in some respects, but not in all. We will be concerned with how many structures (hands) can be built out of given building blocks (cards).

A card $C = C(n, p)$ now has only its weight n and its picture p. For each $n = 1, 2, \ldots$ there is a deck $\mathcal{D}_n$ that contains d_n cards, all of weight n. A hand is a multiset of cards. That is, we may reach into one of the decks $\mathcal{D}_r$ and pull out of it some number of copies of a single card $C(r, p')$, then a number of copies of $C(r, p'')$, and so forth, then from another deck we can take more cards, etc.

No significance attaches to the sequence of cards in the hand. What matters is which cards have been selected and with which multiplicities. The weight of a hand is the sum of the weights of the cards in the hand, taking account of their multiplicities. As before, we let $h(n, k)$ be the number of hands of weight n that contain exactly k cards, and we let

$$\mathcal{H}(x, y) = \sum_{n,k} h(n, k) x^n y^k. \qquad (3.14.1)$$

Notice that the '$n!$' is missing in the assumed form of the generating function. Instead of the mixed egf-ops that was appropriate for labeled counting, a pure ops is the way to go for unlabeled counting.

We need a generic name for the systems that we are constructing. We will call them *prefabs* (instead of exponential families, which applies in the labeled case), and will use letters like $\mathcal{P}$ to represent them.

Thus a prefab $\mathcal{P}$ consists of a sequence of decks $\mathcal{D}_1, \mathcal{D}_2, \ldots$ from which we can form hands, as described above. In $\mathcal{P}$ we let $\mathcal{D}(x) \overset{ops}{\longleftrightarrow} \{d_n\}_1^\infty$.

The main problem is to find the functional relationship between $\mathcal{H}(x, y)$ and $\mathcal{D}(x)$, so let's do that now. We will use the Sorcerer's Apprentice method once more.

For the trickle, consider a prefab $\mathcal{P}$ that consists of just one nonempty deck, $\mathcal{D}_r$, and suppose that $\mathcal{D}_r$ contains only a single card.

In this prefab, a hand H is a fairly simple-minded thing. It consists of some number, k say, of copies of the one and only card that there is, and its weight will be $n = rk$. Hence in this prefab the number $h(n, k)$ of hands of weight n that have exactly k cards is 1 if $n = rk$ and is 0 else. Thus

$$\begin{aligned}
\mathcal{H}(x, y) &= \sum_{n,k} h(n, k) x^n y^k \\
&= \sum_{k \geq 0} 1 \cdot x^{rk} y^k \\
&= \frac{1}{1 - y x^r}.
\end{aligned} \qquad (3.14.2)$$

Next, just as in section 3.4, we define the *merge* operation. If $\mathcal{P}'$ and $\mathcal{P}''$ are prefabs whose picture sets are disjoint, then by their merger $\mathcal{P} = \mathcal{P}' \oplus \mathcal{P}''$ we mean the prefab whose deck $\mathcal{D}_n$, for each n, is the union of the corresponding decks of $\mathcal{P}'$ and $\mathcal{P}''$. If there were d'_n, d''_n cards, respectively, in those two decks, then there are $d_n = d'_n + d''_n$ cards in $\mathcal{D}_n$.

Fundamental lemma of unlabeled counting. *Let $\mathcal{H}'(x, y)$, $\mathcal{H}''(x, y)$ and $\mathcal{H}(x, y)$ be the hand enumerators of prefabs $\mathcal{P}'$, $\mathcal{P}''$ and $\mathcal{P} = \mathcal{P}' \oplus \mathcal{P}''$, respectively. Then $\mathcal{H} = \mathcal{H}'\mathcal{H}''$.*

Proof. Consider a hand $H \in \mathcal{P}$, of weight n, and containing exactly k cards. Some k' of those cards come from $\mathcal{P}'$, and their total weight is, say, n', while the remaining $k - k'$ cards come from $\mathcal{P}''$, and their total weight must be $n - n'$. Thus

$$h(n, k) = \sum_{k', n'} h'(n', k')h''(n - n', k - k'),$$

but, by a strange coincidence, that is exactly the relationship which holds between the coefficients of the power series $\mathcal{H}$, $\mathcal{H}'$ and $\mathcal{H}''$. ∎

Armed with the fundamental lemma, we can now consider a slightly more complicated prefab $\mathcal{P}_r$, which still contains just one nonempty deck $\mathcal{D}_r$, but now that deck contains d_r different cards. By induction on $d_r = 1, 2, \ldots$, we see at once that the hand enumerator of this prefab is

$$\mathcal{H}(x, y) = \frac{1}{(1 - yx^r)^{d_r}}. \tag{3.14.3}$$

Finally (*déjà vu* anybody?), in a general prefab $\mathcal{P}$ in which there are d_n cards in deck $\mathcal{D}_n$, for each $n = 1, 2, 3, \ldots$, we observe that $\mathcal{P} = \oplus_{n=1}^{\infty} \mathcal{P}_n$, where the $\mathcal{P}_n$ are as defined in the previous paragraph. We obtain at once:

Theorem 3.14.1. *In a prefab $\mathcal{P}$ whose hand enumerator is $\mathcal{H}(x, y)$ we have*

$$\mathcal{H}(x, y) = \prod_{n=1}^{\infty} \frac{1}{(1 - yx^n)^{d_n}}, \tag{3.14.4}$$

where d_n is the number of cards in the nth deck $(n \geq 1)$.

This is the analogue of the exponential formula in the case where there are no labels. Like the exponential formula, this one too has an astounding number of elegant applications, and we will discuss a number of them in the sequel. Before we get to that, let's convert (3.14.4) into a formula from which we could actually compute the h's from the d's, using the '$yD\log$' method of section 1.6.

If we take the logarithm of both sides of (3.14.4),

$$\log \mathcal{H}(x, y) = \sum_{s=1}^{\infty} \log \frac{1}{(1 - yx^s)^{d_s}}$$

$$= \sum_{s \geq 1} d_s \log \frac{1}{(1 - yx^s)}$$

$$= \sum_{s \geq 1} d_s \sum_{m \geq 1} \frac{y^m x^{sm}}{m}$$

$$= \sum_{n, m \geq 1} d_{\frac{n}{m}} x^n \frac{y^m}{m},$$

where d_j is to be interpreted as 0 if its subscript is not a positive integer. Next we differentiate with respect to y and multiply by $y\mathcal{H}$, getting

$$y \frac{\partial \mathcal{H}(x, y)}{\partial y} = \mathcal{H}(x, y) \sum_{n, m \geq 1} x^n y^m d_{\frac{n}{m}}.$$

Finally, we take $[x^n y^m]$ of both sides, which yields

$$m h(n, m) = \sum_{r, m' \geq 1} h(n - rm', m - m') d_r \quad (n, m \geq 1; h(n, 0) = \delta_{n,0}).$$

$$(3.14.5)$$

This recurrence holds in any prefab, and permits the numerical computation of the hand counts from the deck counts.

Often the 2-variable deck enumerators $\mathcal{H}(x, y)$ or $\{h(n, k)\}_{n, k \geq 0}$ give more detail than is necessary. If $h_n = \sum_k h(n, k)$ is the number of hands of weight n, however many cards they contain, and if

$$\mathcal{H}(x) \overset{ops}{\longleftrightarrow} \{h_n\}_0^{\infty},$$

then, since we obtain $\mathcal{H}(x)$ from $\mathcal{H}(x, y)$ by formally replacing y by 1, the general counting theorem (3.14.4) becomes

$$\mathcal{H}(x) = \prod_{r=1}^{\infty} \frac{1}{(1 - x^r)^{d_r}}. \qquad (3.14.6)$$

The recurrence (3.14.5) can be replaced by

$$n h_n = \sum_{m \geq 1} D_m h_{n-m} \qquad (n \geq 1; h_0 = 1), \qquad (3.14.7)$$

where $D_m = \sum_{r \backslash m} r d_r$ $(m = 1, 2, \ldots)$.

Considerably more detailed information can be obtained with just a little more effort. Suppose we restrict the *multiplicities* with which the cards can be used in hands. For instance, suppose we decree that every card that appears in a hand must appear there with multiplicity that is divisible by 3, etc. Then what can be said about the number of hands?

Let W be a fixed set of nonnegative integers, containing 0. For each n and k we let $h(n, k; W)$ be the number of hands of weight n that have exactly k cards (counting multiplicities!), each appearing with a multiplicity that belongs to W. Let

$$\mathcal{H}(x, y; W) = \sum_{n,k} h(n, k; W) x^n y^k.$$

Finally, let

$$w(t) = \sum_{k \in W} t^k. \tag{3.14.8}$$

The generating functions are again multiplicative under merger of prefabs with disjoint picture sets. Consider a prefab with just 1 card of weight r, and no other decks. Then $h(n, k; W) = 1$ if $k \in W$ and $n = kr$, and is 0 otherwise, and so

$$\mathcal{H}(x, y; W) = \sum_{k \in W} x^{kr} y^k = w(yx^r).$$

If there are d_r cards in the rth deck, and no other cards, then $\mathcal{H}(x, y; W) = w(yx^r)^{d_r}$, and finally we obtain:

Theorem 3.14.2. *Let the prefab $\mathcal{P}$ contain decks of sizes $d_1, d_2, \ldots$, and let W be a set of nonnegative integers, $0 \in W$. If $h(n, k; W)$ is the number of hands of k cards of weight n, such that each card appears with a multiplicity that belongs to W, then*

$$\mathcal{H}(x, y; W) = \sum_{n,k} h(n, k; W) x^n y^k = \prod_{r \geq 1} w(yx^r)^{d_r}, \tag{3.14.9}$$

where $w(t)$ is given by (3.14.8). ∎

Observe that the theorem reduces to theorem 3.14.1 in the case where $W = Z^+$, the set of all nonnegative integers.

A noteworthy special case is $W = \{0, 1\}$, which means that we can choose a card for our hand or not, but we can't take more than one copy of it. In that case (3.14.9) gives

$$\mathcal{H}(x, y; \{0, 1\}) = \prod_{r \geq 1} (1 + yx^r)^{d_r}$$

$$= \frac{1}{\mathcal{H}(x, -y; Z^+)}. \tag{3.14.10}$$

We proceed with several examples of the use of these formulas.

3.15 The money changing problem

Suppose that in the coinage of a certain country there are 5-cent coins, 11-cent coins, and 37-cent coins. In how many ways can we make change for \$17.19?

In general terms, we are given M positive integers

$$1 \le a_1 < a_2 < \cdots < a_M,$$

and we ask the following question: for each positive integer n, in how many ways can we write

$$n = x_1 a_1 + x_2 a_2 + \cdots + x_M a_M \qquad (\forall i : x_i \ge 0), \qquad (3.15.1)$$

where the x's are integers? This problem is of great importance in a number of areas, both pure and applied, and it has a very beautiful theory, some of which we will give here.

For given $a_1, \ldots, a_M$ we write $S = S(a_1, \ldots, a_M)$ for the set of all n that can be written in the form (3.15.1). S is a semigroup of nonnegative integers.

First let's identify the prefab $\mathcal{P}$ in which everything will be happening. The decks are almost all empty. The only decks that are not empty are the M decks $\mathcal{D}_{a_1}, \ldots, \mathcal{D}_{a_M}$. Each of these contains just a single card. Hence the deck enumerating sequence is

$$d_n = \begin{cases} 1 & \text{if } n = a_1, \ldots, a_M \\ 0 & \text{else.} \end{cases}$$

In a sense, then, the problem is all over. If $h(n, k)$ denotes the number of ways of making change that use exactly k coins, i.e., the number of representations (3.15.1) in which $\sum_i x_i = k$, then according to the main counting theorem (eq. (3.14.4)) we have

$$\mathcal{H}(x, y) = \frac{1}{(1 - yx^{a_1})(1 - yx^{a_2}) \cdots (1 - yx^{a_M})}. \qquad (3.15.2)$$

If h_n is the number of ways of representing n without regard to the number of coins, then from the cruder formula (3.14.6)

$$\mathcal{H}(x) = \frac{1}{(1 - x^{a_1})(1 - x^{a_2}) \cdots (1 - x^{a_M})}. \qquad (3.15.3)$$

Even though the generating functions are known, substantial questions remain. Here are a few of them.

How can we describe the set $\mathcal{S}$? That is, which sums of money can be changed? Given 8-cent and 12-cent coins only, it wouldn't be reasonable to expect to make change for 53 cents. In general, if the greatest common divisor of the set $\{a_1, \ldots, a_M\}$ is $g > 1$, then only multiples of g can be represented. But suppose that $g = 1$, i.e., that the a_i's are *relatively prime*. Then which integers are representable? The central result of this subject is due to I. Schur. It states that $\mathcal{S}$ *then contains all sufficiently large integers, i.e. there exists an integer N such that every integer $n \geq N$ is representable in the form (3.15.1)*.

The smallest integer N that has the property stated in the theorem will be called the *conductor* of the set $\mathcal{S} = \{a_1, \ldots, a_M\}$, and will be denoted by the symbol $\kappa = \kappa(\mathcal{S})$.

For instance, every integer ≥ 8 can be represented as a nonnegative integer linear combination of 3 and 5, and 7 cannot be so represented, so $\kappa(\{3, 5\}) = 8$.

The problem of determining the conductor of a set $\mathcal{S}$ exactly seems to be of enormous difficulty. There are no general 'formulas' for the conductor if $M \geq 3$, and no good algorithms for calculating it if $M \geq 4$. The case $M = 2$ is already very pretty, and the answers are known, so here they are:

Theorem 3.15.1. *Let a and b be relatively prime positive integers. Then*
  (a) every integer $n \geq \kappa = (a-1)(b-1)$ is of the form $n = xa + yb$, $x, y \geq 0$, and
  (b) the integer $\kappa - 1$ is not of that form, and
  (c) of the integers $0, 1, 2, \ldots, \kappa - 1$, exactly half are representable and half are not.

Proof. (Our proof follows [NW]) Since $gcd(a, b) = 1$, we can certainly write every integer m as $xa + yb$ if x, y can have either sign. The representation is unique if we require that $0 \leq x < b$. Then $m \in \mathcal{S}$ if $y \geq 0$, and $m \notin \mathcal{S}$ if $y < 0$. The largest integer that is not representable is therefore obtained by choosing $x = b-1$, $y = -1$. Hence $\kappa(\mathcal{S})$ is one unit larger than $(b-1)a - b$, and parts (a) and (b) of the theorem are proved.

To prove (c), let $0 \leq m < \kappa(\mathcal{S})$, and again consider the unique way of writing $m = xa + yb$, with $0 \leq x < b$. Then

$$m' = \kappa - 1 - m = (b - 1 - x)a + (-1 - y)b.$$

Now $0 \leq b - 1 - x < b$, so if $y \geq 0$ then m is representable and m' is not, while if $y < 0$ then m' is representable and m is not. Hence exactly half of the numbers $0, 1, \ldots, \kappa - 1$ are representable. ∎

Now we're going to prove Schur's theorem. The idea of the proof is that we will consider (without ever writing it down) the partial fraction expansion of the right side of (3.15.3). Among the multitude of terms that occur there we will identify one term whose power series coefficients grow more rapidly than any other, and this will give the desired result.

The generating function $\mathcal{H}(x)$ in (3.15.3) is a rational function whose poles all lie on the unit circle $|x| = 1$. In fact, the poles are at various roots of unity.

What are the multiplicities of these poles? The point $x = 1$ is a pole of multiplicity M, because the denominator of $\mathcal{H}(x)$ is divisible by $(1 - x)^M$. Let $\omega = e^{2\pi i r/s}$ be a primitive (i.e., $\gcd(r,s) = 1$) sth root of 1. What is the multiplicity with which this point $x = \omega$ occurs as a pole of $\mathcal{H}(x)$? It is equal to the number of a_i's that are divisible by s. Since the a_i's are relatively prime, it cannot be that *all* of them are divisible by s.

Therefore $x = 1$ *is a pole of order M of $\mathcal{H}(x)$, and every other pole has multiplicity $< M$.*

Suppose ω is a pole of order r. Then the portion of the partial fraction expansion of $\mathcal{H}$ that comes from ω is of the form

$$\frac{c_1}{(1 - x/\omega)^r} + \frac{c_2}{(1 - x/\omega)^{r-1}} + \cdots .$$

Now refer to the power series expansion (2.5.7), which we repeat here:

$$\frac{1}{(1 - x)^{k+1}} = \sum_{n \geq 0} \binom{n + k}{k} x^n .$$

If $k = 1$, the coefficients of this expansion are linear functions of n. If $k = 2$ they are quadratic functions of n. In general, the coefficients of x^n are growing, as $n \to \infty$, like $n^k/k!$.

The contribution of one fixed pole of order r to the coefficient sequence of $\mathcal{H}(x)$ therefore grows like cn^{r-1}. There is one pole, at $x = 1$, of order M. Its portion of the partial fraction expansion contributes $\sim cn^{M-1}$ to the nth coefficient of $\mathcal{H}(x)$. Since all other poles have strictly lower multiplicities, none of them can alter the asymptotic rate of growth that is contributed by the principal pole at $x = 1$. Hence, for $n \to \infty$ we have $h_n \sim cn^{M-1}$. That certainly implies that for all large enough values of n we will have $h_n \neq 0$, and that finishes the proof. However, as long as we're here, why not find out the value of c also?

The partial fraction expansion of $\mathcal{H}(x)$ is of the form

$$\mathcal{H}(x) = \frac{1}{(1 - x^{a_1})(1 - x^{a_2}) \cdots (1 - x^{a_M})}$$

$$= \frac{c}{(1 - x)^M} + O((1 - x)^{-M+1}).$$

To calculate c, multiply both sides by $(1 - x)^M$ and let $x \to 1$. This gives $c = 1/(a_1 \cdots a_M)$. Thus we get a growth estimate along with the proof of the theorem.

Theorem 3.15.2 (Schur's theorem). *If h_n denotes the number of representations of n as a nonnegative integer linear combination of $a_1, \ldots, a_M$, these being a relatively prime set of positive integers, then*

$$h_n \sim \frac{n^{M-1}}{(M-1)!a_1 a_2 \cdots a_M} \qquad (n \to \infty). \qquad (3.15.4)$$

In particular, there exists an integer N such that every $n \geq N$ is so representable in at least one way. ∎

Example 1.

Given two relatively prime integers a, b. Find an explicit formula for $f(n)$, the number of ways to change n cents using those coins.

From (3.15.3) we have

$$\sum_n f(n)x^n = \frac{1}{(1-x^a)(1-x^b)}, \qquad (3.15.5)$$

so what remains is a partial fraction expansion. We find

$$\frac{1}{(1-x^a)(1-x^b)} = \frac{A}{(1-x)^2} + \frac{B}{(1-x)} + \sum_{\substack{\omega^a=1 \\ \omega \neq 1}} \frac{C_\omega}{1-x/\omega} + \sum_{\substack{\zeta^b=1 \\ \zeta \neq 1}} \frac{D_\zeta}{1-x/\zeta}. \qquad (3.15.6)$$

As regards the constants, we already know that $A = 1/(ab)$, from (3.15.4). To find B, multiply (3.15.6) by $(1-x)^2$, differentiate, and let $x - 1$. This gives $B = (a+b-2)/(2ab)$. To find C_ω, multiply by $(1-x/\omega)$ and let $x = \omega$. The result is that $C_\omega = 1/(a(1-\omega^b))$, and similarly for D_ζ. Finally we take the coefficient of x^n throughout (3.15.6) to get the formula

$$f(n) = \frac{n}{ab} + \frac{a+b}{2ab} + \sum_{\substack{\omega^a=1 \\ \omega \neq 1}} \frac{C_\omega}{\omega^n} + \sum_{\substack{\zeta^b=1 \\ \zeta \neq 1}} \frac{D_\zeta}{\zeta^n}. \qquad (3.15.7)$$

If we examine the two sums that appear in (3.15.7) as functions of n, we see that each of them is a periodic function of n. The first sum is periodic of period a and the second is periodic of period b. The sum of these two sums is therefore periodic of period ab. We have therefore found that *the number of ways to change n cents into coins of a- and b-cent denominations is*

$$f(n) = \frac{n}{ab} + \frac{a+b}{2ab} + per(n), \qquad (3.15.8)$$

where $per(n)$ is periodic of period ab, and is on the average 0.

We might like to see this periodicity in action, so let's take $a = 3$ and $b = 5$. A good way to compute the numbers $f(n)$ is to use the recurrence formula that is implicit in the generating function (3.15.5). If we use the $xD\log$ method on (3.15.5), we find the recurrence in the form

$$nf(n) = 3\sum_{j\geq 1} f(n-3j) + 5\sum_{j\geq 1} f(n-5j) \qquad (n \geq 1; f(0) = 1), \quad (3.15.9)$$

with the understanding that $f(m) = 0$ if $m < 0$. Table 3.1 shows n, $f(n)$, and $15(f(n) - (n/15) - (4/15))$ (which is periodic of period 15, according to (3.15.8)). ∎

0	1	2	3	4	5	6	7	8	9	10	11	12	13	14
1	0	0	1	0	1	1	0	1	1	1	1	1	1	1
11	−5	−6	8	−8	6	5	−11	3	2	1	0	−1	−2	−3

15	16	17	18	19	20	21	22	23	24	25	26	27	28	29
2	1	1	2	1	2	2	1	2	2	2	2	2	2	2
11	−5	−6	8	−8	6	5	−11	3	2	1	0	−1	−2	−3

Table 3.1

3.16 Partitions of integers

A partition of a positive integer n is a representation

$$n = r_1 + r_2 + \cdots + r_k \qquad (r_1 \geq r_2 \geq \cdots \geq r_k \geq 1). \qquad (3.16.1)$$

The numbers $r_1, \ldots, r_k$ are the *parts* of the partition. Hence (3.16.1) is a partition of n into k parts.

There are 7 partitions of 5, namely 5=5, =4+1, =3+2, =3+1+1, =2+2+1, =2+1+1+1, =1+1+1+1+1. The number of partitions of n is denoted by $p(n)$, and $p(n, k)$ is the number of partitions of n into k parts. The investigation of the deeper properties of $p(n)$ was one of the jewels of 20th century analysis, involving researches of Hardy and Ramanujan and further work by Rademacher, the result of which was an exact closed formula for $p(n)$ that was at the same time a complete asymptotic series. The whole story can be found in Andrews [An].

From our point of view, the theory of partitions is the case of the money-changing problem where coins of every positive integer size are available. Thus, theorem 3.14.1 gives us immediately an opsgf of the partition function in the form of the reciprocal of an infinite product,

$$\sum_{n,k\geq 0} p(n,k)x^n y^k = \frac{1}{(1-yx)(1-yx^2)(1-yx^3)(1-yx^4)\cdots} \qquad (3.16.2)$$

$$(p(0,k) = \delta_{0,k}).$$

With $y = 1$ we find

$$\sum_{n \geq 0} p(n)x^n = \frac{1}{(1 - x)(1 - x^2)(1 - x^3)(1 - x^4)\cdots} \qquad (p(0) = 1) \quad (3.16.3)$$

as the generating function for $\{p(n)\}$ itself.

At the other extreme, we can think about partitions with constrained parts and constrained multiplicities of parts. Let two sets W, of nonnegative integers, and R, of positive integers, be given, with $0 \in W$. Let $p(n, k; W, R)$ be the number of partitions of n into k parts such that all of the parts lie in R, and all of their multiplicities lie in W. Then from (3.14.9)

$$\sum_{n,k} p(n, k; W, R)x^n y^k = \prod_{r \in R} \left(\sum_{k \in W} y^k x^{kr} \right). \qquad (3.16.4)$$

From this generating function we can prove many theorems about partitions.

Example 1.

Let $W - \{0, 1\}$, $R - \{1, 2, \ldots\}$. Then $p(n, k; W, R)$ is the number of partitions of n into k *distinct* parts, and we have

$$\sum_{n,k} p(n, k; \{0, 1\}, \mathbf{Z}^+)x^n y^k = \prod_{r \geq 1}(1 + yx^r). \qquad (3.16.5)$$

If we let $y = 1$ we obtain

$$\sum_n p(n; \{0, 1\}, \mathbf{Z}^+)x^n = \prod_{r \geq 1}(1 + x^r)$$

$$= \prod_{r \geq 1} \frac{1 - x^{2r}}{1 - x^r}$$

$$= \frac{(1 - x^2)(1 - x^4)\cdots}{(1 - x)(1 - x^2)(1 - x^3)(1 - x^4)\cdots}$$

$$= \frac{1}{(1 - x)(1 - x^3)(1 - x^5)(1 - x^7)\cdots}.$$

The last member, however, generates the partitions of n into odd parts, and we have a generating function proof of:

Theorem 3.16.1. *For each* $n = 1, 2, 3, \ldots$*, the number of partitions of* n *into odd parts is equal to the number of partitions of* n *into distinct parts.*
∎

For instance, the partitions of 5 into odd parts are 5, 3+1+1, and 1+1+1+1+1, while its partitions into distinct parts are 5, 4+1, and 3+2.

Theorem 3.16.1 was discovered by Euler. A great many proofs of it have been given. Some of the most interesting proofs are *bijective*; that is, they give explicit constructions that match each partition into odd parts with a partition into distinct parts. ∎

Example 2.

Now let $W = \{0, 1, \ldots, q\}$ and $R = \mathbf{Z}^+$. The right side of (3.16.4) becomes

$$\prod_{r \geq 1}(1 + t^r + \cdots + t^{qr}) = \prod_{r \geq 1}\left(\frac{1 - t^{r(q+1)}}{1 - t^r}\right).$$

Each factor in the numerator of this product cancels one in the denominator, leaving in the denominator only those factors in which r is not divisible by $q + 1$. This proves the following result, which reduces to theorem 3.16.1 when $q = 1$.

Theorem 3.16.2. *Fix $q \geq 1$. For each $n \geq 1$, the number of partitions of n into parts that are not divisible by $q + 1$ is equal to the number of partitions of n in which no part appears more than q times.* ∎

3.17 Rooted trees and forests

A rooted tree is a tree whose vertices are unlabeled, except that one of them is distinguished as 'the root.'

In section 3.12 we counted labeled trees, using the exponential formula. Here we will count unlabeled, rooted trees. On each card in a deck $\mathcal{D}_n$ there is now the integer n and a picture of a rooted tree of n vertices. The deck $\mathcal{D}_4$ is shown in Fig. 3.5.

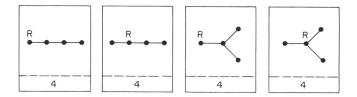

Fig. 3.5: The rooted trees of 4 vertices

A hand of weight n and k cards is, in this case, a rooted forest of n vertices and k connected components (rooted trees). If $h(n, k)$ is the number of these and if $h(n) = \sum_k h(n, k)$ is the number of all rooted forests of n vertices, then by (3.14.6)

$$\sum_n h(n)x^n = \prod_{n \geq 1} \frac{1}{(1 - x^n)^{t(n)}}, \qquad (3.17.1)$$

where $t(n) = h(n,1)$ is the number of rooted trees of n vertices.

Next, just as we found in the labeled case (see 3.12.2), there is a simple relationship between the number of rooted forests of n vertices and of rooted trees of $n+1$ vertices: they are equal. Just add a new vertex r to the forest, call it the new root, connect it to all of the former roots of the trees in the forest, and there is the rooted tree that corresponds to the given forest.

Hence $h(n) = t(n+1)$ $(n \geq 0)$. Then (3.17.1) takes the form

$$\sum_n t(n+1)x^n = \prod_{n \geq 1} \frac{1}{(1-x^n)^{t(n)}}. \tag{3.17.2}$$

This equation in fact determines all of the numbers $\{t(n)\}$. It is, however, a fairly formidable equation, and we should not expect simple formulas for these numbers.

3.18 Historical notes

The exponential formula first appeared in the thesis of Riddell [RU], in the form of counting connected labeled graphs from a knowledge of the number of all labeled graphs. Since then the idea has been generalized and extended by several researchers.

In [BG], and at about the same time in [FS], significant extensions of the idea were made to very general labeled and unlabeled applications, by Bender and Goldman and by Foata and Schützenberger. The former introduced 'prefabs' and the latter used the '*composé partitionnel.*' Further developments of the method can be found in Stanley ([St1], [St2]) who worked with a partition-based approach, in Joyal [Jo] who used functorial methods in his theory of 'species,' in Beissinger [Bei], and in Garsia and Joni [GaJ].

The approach taken in this book is most closely akin to the *composé partitionnel.* The suggestion to cast the discussion in terms of cards, decks, and hands was made to me in private conversation by Adriano Garsia, when I showed him a set of lecture notes of mine that were based on the graph-theoretical point of view. I think that his suggestion affords maximum clarity of the ideas along with maximum generality of applications.

Exercises

1. Give an explicit 1-1 correspondence between partitions of n into distinct parts and partitions of n into odd parts.

2. Fix integers n, k. Let $f(n, k)$ be the number of permutations of n letters whose cycle lengths are all divisible by k. Find a simple, explicit egf for $\{f(n, k)\}_{n \geq 0}$. Find a simple, explicit formula for $f(n, k)$. (Hint: You might need the discussion at the end of section 3.4.)

3. Find the egf for the partitions of the set $[n]$, all of whose classes have a prime number of elements.

4. In a group Γ, the *order* of an element g is the least positive integer ρ such that $g^\rho = 1_\Gamma$.

    (a) In the group of all permutations of n letters, express the order of a permutation σ in terms of the lengths of its cycles.

    (b) Let $g(n, k)$ be the number of permutations of n letters whose order is k. Express $g(n, k)$ in terms of the number $\tilde{g}(n, m)$ of n-permutations whose cycle lengths all divide m.

5. Let T_n be the number of involutions of n letters.

    (a) Find a recurrence formula that is satisfied by these numbers.

    (b) Compute $T_1, \ldots, T_6$.

    (c) Give a combinatorial and constructive interpretation of the recurrence. That is, after having derived it from the generating function, re-derive it without the generating function.

6. Find, in simple form, the egf of the sequence of numbers of permutations of n letters that have no cycles of lengths ≤ 3. Your answer should not contain any infinite series.

7. Find the generating function for labeled graphs with all vertices of degrees 1 or 2, and an odd number of connected components. Find a recurrence formula for these numbers, calculate the first few, and draw the graphs involved.

8. From (3.5.2) find a three term recurrence relation that is satified by the Stirling numbers of the first kind. Give a direct combinatorial proof of this recurrence relation. That is, reprove it, without using any generating functions.

9. As in section 3.7, find the egf of the numbers $\{g(n)\}_0^\infty$ of permutations of n letters that have both of the following two properties: (a) they have an odd number of cycles and (b) the lengths of all of their cycles are even. Find a simple, explicit formula for these numbers.

10. Find an explicit formula for $\left\{{n \atop k}\right\}$, the Stirling number of the second kind by expanding the kth power that appears in (3.6.2) by the binomial

theorem. Your formula should be in the form of a single finite sum.

11. Let S, T be fixed sets of positive integers. Let $f(n; S, T)$ be the number of partitions of $[n]$ whose class sizes all lie in S and whose number of classes lies in T. Show that $\{f(n; S, T)\}_{n \geq 0}$ has the egf $e_T(e_S(x))$, where $e_S(x) = \sum_{s \in S} x^s / s!$.

12. Fix $k > 0$. Let $f(n, k)$ be the number of permutations of n letters whose longest cycle has length k. Find the egf of $\{f(n, k)\}_{n \geq 0}$, for k fixed.

13. If $T(x)$ and $\mathcal{G}(x)$ denote, respectively, the egf's of involutions, in (3.8.3), and of 2-regular graphs, in (3.9.1), then observe that

$$T(x)\mathcal{G}(x)^2 = \frac{1}{1 - x}.$$

(a) Write out the identity between the sequences $\{g(n)\}$, $\{t_n\}$ that is implied by the above generating function relation.

(b) Show that for each fixed $n \geq 1$ there are exactly the same numbers of

    (i) permutations of n letters and of

    (ii) triples (τ, G_1, G_2), where τ is an involution of a set R, G_1 is a 2-regular graph on a vertex set S, G_2 is a 2-regular graph on a vertex set T, and R, S, T partition $[n]$.

(c) Find, explicitly, a 1-1 correspondence such as is described in part (b) above.

14. Let $\mathcal{F}$ be an exponential family with associated polynomials $\{\phi_n(x)\}$ of binomial type, and with deck enumerator $\mathcal{D}(x)$.

(a) If D_y denotes the differential operator $\partial/\partial y$, then show that

$$\mathcal{D}^{(-1)}(D_y)\phi_n(y) = n\phi_{n-1}(y) \qquad (n \geq 0)$$

by directly applying the operator to the egf of the polynomial sequence (here $\mathcal{D}^{(-1)}$ denotes the inverse function in the sense of functional composition).

(b) In the case of the exponential family of permutations by cycles, find the associated polynomials of binomial type, and verify the identity proved in part (a) by direct computation with those polynomials.

15. In an exponential family $\mathcal{F}$, let $\tilde{h}(n)$ be the number of hands of weight n whose cards have all different weights.

(a) Show that

$$\sum_{n \geq 0} \frac{\tilde{h}(n)}{n!} x^n = \prod_{k=1}^{\infty} \left\{ 1 + \frac{d_k}{k!} x^k \right\}.$$

(b) Let p_n be the probability that a permutation of n letters has cycles whose lengths are all different. Then

$$\{p_n\} \overset{ops}{\longrightarrow} \prod_{k\geq 1}\left\{1 + \frac{x^k}{k}\right\}.$$

(c) If $p(x)$ denotes the generating function in part (b) above, determine the growth of $p(x)$ as $x \to 1^-$. Do this by inserting additional factors of $e^{-x^k/k}$ in the product.

16. Let numbers $\{c_n\}$ be defined by

$$x^x = 1 + \sum_{n\geq 1} \frac{c_n}{n!}(x-1)^n.$$

Show that each c_n is an integer multiple of n, and in fact is a multiple of $n(n-1)$ if and only if $n-1$ divides $(n-2)!$.

17. Here we want to show that the Stirling numbers of the first and second kinds are inverse to each other, in a certain sense. In the generating function (3.5.2) for the former, replace x by $1/x$ and compare with the generating function (1.6.5) for the latter. Multiply the functions together so that the hard part cancels out. Read off the coefficient of x^n in what remains, and state it as an assertion that a certain pair of matrices, each involving Stirling numbers, are inverses of each other.

18. Let a_n be the number of *unlabeled* graphs of n vertices each of whose connected components is a path or a cycle. Let $F(x)$ be the opsgf of the sequence $\{a_n\}$. Find $F(x)$ and express it in terms of Euler's opsgf for the sequence $\{p(n)\}$ of the numbers of partitions of integers n.

19. Let a_n be the number of unlabeled rooted trees of n vertices in which the degree of the root is 2. That is, there are exactly 2 edges incident at the root. Let $T(x)$ be the opsgf of the sequence $\{t_n\}$ that counts *all* rooted trees of n vertices. Show that

$$\sum_n a_n x^{n-1} = \frac{1}{2}\left(T(x)^2 + T(x^2)\right).$$

20. Find the largest integer that is *not* of the form $6x + 10y + 15z$ where x, y, z are nonnegative integers. *Prove* that your answer is correct, i.e., that your integer is not so representable, and that every integer larger than it is so representable.

21. In a country that has 1-cent, 2-cent, and 3-cent coins only, the number of ways of changing n cents is exactly the integer *nearest* to $(n+3)^2/12$.

22. This exercise develops a considerable sharpening of the exponential formula, that will be used again in section 4.7.

(a) In an exponential family $\mathcal{F}$, the number of hands of weight n that contain exactly a_1 cards of weight 1 and a_2 cards of weight 2 and a_3 of weight 3 and $\ldots$, where $a_1 + 2a_2 + \cdots = n$, is the coefficient of $(t^n x_1^{a_1} x_2^{a_2} \cdots)/n!$ in the expansion of

$$\exp \left\{ \sum_{i \geq 1} \frac{x_i d_i t^i}{i!} \right\}.$$

(b) Let $f(n, r, s)$ be the number of partitions of the set $[n]$ that have exactly r classes of size 1 and exactly s classes of size 2 (however many classes of other sizes they may have). Then

$$\sum_{n,r,s} f(n, r, s) x^r y^s \frac{t^n}{n!} = \exp \left(xt + \frac{yt^2}{2} + e^t - 1 - t - \frac{t^2}{2} \right).$$

Chapter 4
Applications of generating functions

4.1 Generating functions find averages, etc.

Power series generating functions are exceptionally well adapted to finding means, standard deviations, and other moments of distributions, with minimum work. Suppose $f(n)$ is the number of objects, in a certain set S of N objects, that have exactly n properties, for each $n = 0, 1, 2, \ldots$, with $\sum_n f(n) = N$. What is the average number of properties that an object in S has? Evidently it is

$$\mu = \frac{1}{N} \sum_n n f(n). \qquad (4.1.1)$$

Suppose we happen to be fortunate enough to be in possession of the opsgf of the sequence $\{f(n)\}$, say $F(x) \overset{ops}{\longleftrightarrow} \{f(n)\}$. Is there some convenient way to express the mean μ of (4.1.1) in terms of F? But of course. Clearly, $\mu = F'(1)/F(1)$. So averages can be computed directly from generating functions.

Let's go to the next moment, the standard deviation σ, of the distribution. This is defined as follows:

$$\sigma^2 = \frac{1}{N} \sum_{\omega \in S} (n(\omega) - \mu)^2, \qquad (4.1.2)$$

where ω represents an object in the set S, and $n(\omega)$ is the number of properties that ω has. σ^2, which is known as the *variance* of the distribution, is therefore the mean square of the difference between the number of properties that each object has and the mean number of properties μ.

Every one of the $f(n)$ objects ω that has exactly n properties will contribute $(n - \mu)^2$ to the sum in (4.1.2), and therefore

$$
\begin{aligned}
\sigma^2 &= \frac{1}{N} \sum_n (n - \mu)^2 f(n) \\
&= \frac{1}{N} \sum_n (n^2 - 2\mu n + \mu^2) f(n) \\
&= \frac{1}{N} \{(xD)^2 - 2\mu(xD) + \mu^2\} F(x)|_{x=1} \\
&= (F''(1) + (1 - 2\mu)F'(1) + \mu^2 F(1))/F(1) \\
&= F''(1)/F(1) + F'(1)/F(1) - (F'(1)/F(1))^2 \\
&= \{(\log F)' + (\log F)''\}_{x=1}.
\end{aligned}
\qquad (4.1.3)
$$

So the standard deviation can also be calculated in terms of the values of F and its first two derivatives at $x = 1$.

Let's work this out in exponential families. In an exponential family $\mathcal{F}$, what is the average number, $\mu(n)$, of cards in a hand of weight n?

If $h(n, k)$ is the number of hands of weight n that have k cards, then the average is

$$\mu(n) = \frac{1}{h(n)} \sum_k kh(n, k). \tag{4.1.4}$$

Now if we begin with the exponential formula

$$\sum_{n,k} h(n, k) \frac{x^n}{n!} y^k = e^{y\mathcal{D}(x)}$$

the thing to do is to apply the operator $\partial/\partial y$ and then set $y = 1$. The result is that

$$\sum_n \frac{x^n}{n!} \sum_k kh(n, k) = \mathcal{D}(x) e^{\mathcal{D}(x)} = \mathcal{D}(x)\mathcal{H}(x). \tag{4.1.5}$$

Theorem 4.1.1. *In an exponential family $\mathcal{F}$, the average number of cards in hands of weight n is*

$$\mu(n) = \left[\frac{h(n)x^n}{n!} \right] \mathcal{D}(x)\mathcal{H}(x)$$
$$= \frac{1}{h(n)} \sum_r \binom{n}{r} d_r h(n - r). \tag{4.1.6}$$

Example 1. Cycles of permutations

The averaging relations (4.1.6) are particularly happy if $h(n) = n!$, as in the family of all permutations. There, (4.1.6) becomes

$$\mu(n) = \frac{1}{n!} \sum_r \binom{n}{r} (r - 1)!(n - r)!$$
$$= 1 + \frac{1}{2} + \frac{1}{3} + \cdots + \frac{1}{n}.$$

Consequently, the average number of cycles in a permutation of n letters is the harmonic number H_n.

What is the standard deviation? The function $F(x)$ that appears in (4.1.3), in the case of permutations, is, for n fixed,

$$F(x) = \sum_k h(n, k)x^k = x(x + 1)(x + 2) \cdots (x + n - 1),$$

by (3.5.2). After taking logarithms and differentiating, following (4.1.3), we find $F(1) = n!$, $(\log F)'(1) = H_n$, and

$$(\log F)''(1) = -1 - 1/4 - 1/9 - 1/16 - \cdots - 1/n^2.$$

If we substitute this into (4.1.3), we find that the variance of the distribution of cycles over permutations of n letters is

$$\sigma^2 = H_n - 1 - 1/4 - 1/9 - \cdots - 1/n^2$$
$$= \log n + \gamma - \pi^2/6 + o(1).$$

where γ is Euler's constant.

Hence the average number of cycles is $\sim \log n$ with a standard deviation $\sigma \sim \sqrt{\log n}$. ∎

4.2 A generatingfunctionological view of the sieve method

The sieve method* is one of the most powerful general tools in combinatorics. It is explained in most texts in discrete mathematics, however it most often appears as a sequence of manipulations of alternating sums of binomial coefficients. Here we will emphasize the fact that generating functions can greatly simplify the lives of users of the method.

We are given a finite set Ω of objects and a set P of properties that the objects may or may not possess.** In this context, we want to answer questions of the following kind: how many objects have no properties at all? how many have exactly r properties? what is the average number of properties that objects have? etc., etc.

The characteristic flavor of problems that the sieve method can handle is that, although it is hard to see how many objects have *exactly* r properties, for instance, it is relatively easy to see how many objects have *at least* a certain set of properties and maybe more.

What the method does is to convert the 'at least' information into the 'exactly' information.

To see how this works, if $S \subseteq P$ is a set of properties, let $N(\supseteq S)$ be the number of objects that have *at least* the properties in S. That is, $N(\supseteq S)$ is the number of objects whose set of properties contains S.

For fixed $r \geq 0$, consider the sum

$$N_r = \sum_{|S|=r} N(\supseteq S). \qquad (4.2.1)$$

---

* A.k.a. 'the principle of inclusion-exclusion,' and often abbreviated as 'p.i.e.'

** Strictly speaking, a property is just a subset of the objects, but in practice we will usually have simple verbal descriptions of the properties.

Introduce the symbol $P(\omega)$ for the set of properties that ω has. Then we can write N_r as follows:

$$
\begin{aligned}
N_r &= \sum_{|S|=r} N(\supseteq S) \\
&= \sum_{|S|=r} \sum_{\substack{\omega \in \Omega \\ S \subseteq P(\omega)}} 1 \\
&= \sum_{\omega \in \Omega} \left\{ \sum_{\substack{|S|=r \\ S \subseteq P(\omega)}} 1 \right\} \\
&= \sum_{\omega \in \Omega} \binom{|P(\omega)|}{r}.
\end{aligned}
\tag{4.2.2}
$$

Therefore *every object that has exactly t properties contributes $\binom{t}{r}$ to N_r.* If there are e_t objects that have exactly t properties, then (4.2.2) simplifies to

$$
N_r = \sum_{t \geq 0} \binom{t}{r} e_t \qquad (r = 0, 1, 2, \ldots). \tag{4.2.3}
$$

Recall the philosophy of the method: the N_r's are easier to calculate than the e_r's because they can be found from (4.2.1). However, the e_r's are what we want. Therefore it is desirable to be able to solve the equations (4.2.3) for the e's in terms of the N's. But how can we do that? After all, (4.2.3) is a set of simultaneous equations.

At first glance that might seem to be a tall order, but with a friendly generating function at your side, it's easy. Let $N(x)$ and $E(x)$ denote* the opsgf's of the sequences $\{N_r\}$, $\{e_r\}$, respectively. What relation between the two generating functions is implied by the equations (4.2.3)?

Multiply (4.2.3) by x^r and sum on r. We then get

$$
\begin{aligned}
N(x) &= \sum_r \sum_t \binom{t}{r} e_t x^r \\
&= \sum_t e_t \left\{ \sum_r \binom{t}{r} x^r \right\} \\
&= \sum_t e_t (x+1)^t \\
&= E(x+1).
\end{aligned}
\tag{4.2.4}
$$

---

* The letters 'N' and 'E' are intended to suggest the N_r's and the word 'Exactly.'

In the language of generating functions, the set of equations (4.2.3) boils down to the fact that $N(x) = E(x+1)$. Now the problem of solving for the e's in terms of the N's is a triviality, and the solution is obviously

$$E(x) = N(x - 1)$$

$$(4.2.5)$$

This is the sieve method. *The act of replacing the variable x by $x - 1$ in the generating function $N(x)$ replaces the unfiltered data $\{N_r\}$ by the sieved quantities $\{e_r\}$.*

If the N's are known, then in principle we can read off the e's as the coefficients of $N(x-1)$.

For example, e_0 is the number of objects that have no properties at all. By (4.2.5),

$$e_0 = E(0) = N(-1) = \sum_t (-1)^t N_t. \qquad (4.2.6)$$

It's easy to find explicit formulas for all of the e_j's by looking at the coefficient of x^j on both sides of (4.2.5). The result is

$$e_j = \sum_t (-1)^{t-j} \binom{t}{j} N_t. \qquad (4.2.7)$$

But (4.2.5) says it all, in a much cleaner fashion.

We will now summarize the sieve method, and then give a number of examples of its use.

The Sieve Method

(A) (*Find Ω and P*) Given an enumeration problem, find a set of objects and properties such that the problem would be solved if we knew the number of objects with each number of properties.

(B) (*Find the unfiltered counts $N(\supseteq S)$*) For each set S of properties, find $N(\supseteq S)$, the number of objects whose set of properties contains S.

(C) (*Find the coefficients N_r*) For each $r \geq 0$, calculate the N_r by summing the $N(\supseteq S)$ over all sets S of r properties, as in (4.2.1).

(D) (*The answer is here.*) The numbers e_r are the coefficients of the powers of x in the polynomial $N(x-1)$. ∎

Before we get to some examples, we would like to point out that the number N_1 has a special role to play. According to (4.2.3), $N_1 = \sum_t t e_t$. That, however, is what you would want to know if you were trying to calculate the average number of properties that objects have. Hence it is good to remember that *when using the sieve method on a set of N objects, the average number of properties that an object has is N_1/N.*

Example 1. The fixed points of permutations.

Of the $n!$ permutations of n letters, how many have exactly r fixed points?

Step (A) of the sieve method asks us to say what the set of objects is and what the set of properties is. It is almost always worthwhile to be quite explicit about these. In the case at hand, the set Ω of objects is the set of all permutations of n letters. There are n properties: for each $i = 1, \ldots, n$, a permutation τ has property i if i is a fixed point of τ, i.e., if $\tau(i) = i$.

With those definitions of Ω and P, it is indeed true that we would like to know the numbers of objects that have exactly r properties, for each r.

In step (B) we must find the $N(\supseteq S)$. Hence let S be a set of properties. Then $S \subseteq [n]$ is a set of letters, and we want to know the number of permutations of n letters that leave *at least* the letters in S fixed.

If a permutation leaves the letters in S fixed, then it can act freely on only the remaining $n - |S|$ letters, and so there are $(n - |S|)!$ such permutations. Hence

$$N(\supseteq S) = (n - |S|)!.$$

For step (C) we calculate the N_r's. But, for each $r = 0, \ldots, n$,

$$N_r = \sum_{|S|=r} N(\supseteq S) = \sum_{|S|=r} (n - |S|)! = \binom{n}{r}(n - r)! = \frac{n!}{r!}.$$

In step (D) we're ready for the answers. It will save some writing if we introduce the abbreviation $\exp_{|\alpha}$ for the truncated exponential series

$$\exp_{|\alpha}(x) = \sum_{0 \le r \le \alpha} \frac{x^r}{r!}. \tag{4.2.8}$$

Now we form the opsgf $N(x)$ from the N_r's that we just found:

$$N(x) = \sum_{r=0}^{n} \frac{n!}{r!}x^r = n! \sum_{r=0}^{n} \frac{x^r}{r!}.$$

Then e_t is the coefficient of x^t in $N(x-1)$, i.e.,

$$E(x) = \sum_{t} e_t x^t = n! \sum_{r=0}^{n} \frac{(x-1)^r}{r!} = n! \exp_{|n}(x - 1). \tag{4.2.9}$$

As an extra dividend, the *average* number of fixed points that permutations of n letters have is

$$\frac{N_1}{N} = \frac{n!}{n!} = 1.$$

On the average, a permutation has 1 fixed point.

The number of permutations that have no fixed points at all is

$$e_0 = E(0) = N(-1) = n!\,\exp_{|n}(-1) \sim \frac{n!}{e}. \qquad (4.2.10)$$

Finally, if we really want a formula for the e_t's, it's quite easy to find from (4.2.9) that

$$e_t = \frac{n!}{t!}\,\exp_{|(n-t)}(-1)$$
$$\sim e^{-1}\frac{n!}{t!} \qquad (n \to \infty). \qquad (4.2.11)$$

■

Example 2. The number of k-cycles in permutations.

Fix positive integers n, k, and $r \geq 0$. How many permutations of n letters have exactly r cycles of length k?

Whatever the answer is, it should at least have the good manners to reduce to the answer of the previous example when $k = 1$, since a fixed point is a cycle of length 1.

What are the objects and the properties? Evidently Ω is the set of all permutations of n letters. Further, the set P of properties is the set of all possible k-cycles chosen from n letters. How many such k-cycles are there? The k letters can be chosen in $\binom{n}{k}$ ways, and they can be arranged around a cycle in $(k-1)!$ ways, so we are facing a list of $\binom{n}{k}(k-1)!$ properties.

Choose a set S of k-cycles from P. How many permutations have at least the set S of properties? None at all, unless the sets of letters in those cycles are pairwise disjoint. If the sets are pairwise disjoint, then there are $N(\supseteq S) = (n - k|S|)!$ permutations that have at least all of those k-cycles.

Next we calculate N_r, the sum of $N(\supseteq S)$ over all sets of r properties. The terms in this sum are either 0 or $(n - kr)!$. So we really need to know only how many of them are not 0, that is, in how many ways we can choose a set of r k-cycles from n letters in such a way that the cycles operate on disjoint sets of letters.

The letters for the first cycle can be chosen in $\binom{n}{k}$ ways, and they can be ordered around the cycle in $(k-1)!$ ways. The letters for the second cycle can then be chosen in $\binom{n-k}{k}$ ways, and ordered in $(k-1)!$ ways, etc. Finally, since the sequence in which the cycles are constructed is of no significance, we divide by $r!$. Hence

$$N_r = \frac{(n-kr)!}{r!}\,\frac{n!(k-1)!^r}{(k!)^r(n-kr)!}$$
$$= \frac{n!}{k^r r!} \qquad (0 \leq r \leq n/k). \qquad (4.2.12)$$

We can get a little piece of the solution right here, with no more work: the *average* number of k-cycles that permutations of n letters have is $N_1/n! = 1/k$.

The opsgf of $\{N_r\}$ is

$$N(x) = n! \sum_{0 \le r \le n/k} \frac{x^r}{k^r r!}$$

$$= n! \exp_{|(n/k)} \left(\frac{x}{k} \right). \tag{4.2.13}$$

Finally, in the sieving step, we convert this to exact information by replacing x by $x - 1$, to obtain

$$E(x) = n! \exp_{|(n/k)} \left(\frac{x - 1}{k} \right). \tag{4.2.14}$$

Example 3. Stirling numbers of the second kind.

The Stirling numbers $\left\{ {n \atop k} \right\}$, which we studied in section 1.6, are the numbers of partitions of a set of n elements into k classes. We can find out about them with the sieve method if we can invent a suitable collection of objects and properties. For the set Ω of objects we take the collection of all k^n ways of arranging n labeled balls in k labeled boxes. Further, such an arrangement will have property P_i if box i is empty $(i = 1, \ldots, k)$. Then $k! \left\{ {n \atop k} \right\}$ is the number of objects that have exactly no properties.

Let S be some set of properties. How many arrangements of balls in boxes have at least the set S of properties? If $N(\supseteq S)$ is that number, then $N(\supseteq S)$ counts the arrangements of n labeled balls into just $k - |S|$ labeled boxes, because all of the boxes that are labeled by S must be empty.

There are obviously $(k - |S|)^n$ such arrangements. Hence

$$N(\supseteq S) = \begin{cases} (k - |S|)^n & \text{if } |S| \le k, \\ 0, & \text{else.} \end{cases}$$

If we now sum over all sets S of r properties, we obtain for $r \le k$,

$$N_r = \binom{k}{r} (k - r)^n,$$

whose opsgf is

$$N(x) = \sum_{0 \le r \le k} \binom{k}{r} (k - r)^n x^r.$$

We can now invoke the sieve to find that the number of arrangements that have exactly t empty cells is the coefficient of x^t in $N(x - 1)$. On the other hand, the number of arrangements that have exactly t empty cells is clearly

$$\binom{k}{t}(k-t)!\left\{{n \atop k-t}\right\} = \frac{k!}{t!}\left\{{n \atop k-t}\right\}.$$

The result is the identity

$$\sum_{0\leq r\leq k}\binom{k}{r}(k-r)^n(x-1)^r = k!\sum_{0\leq t\leq k}\left\{{n \atop k-t}\right\}\frac{x^t}{t!}. \qquad (4.2.15)$$

If we put $x = 0$, we find the explicit formula (1.6.7) again.

If, on the other hand, we compare (4.2.15) with the rule (2.3.3) for finding the coefficients of the product of two egf's, we discover the following remarkable identity:

$$\sum_{1\leq k\leq n}\left\{{n \atop k}\right\}y^k = e^{-y}\sum_{r\geq 1}\frac{r^n}{r!}y^r. \qquad (4.2.16)$$

This shows that e^{-y} times the infinite series is a polynomial! The special case $y = 1$ has been previously noted in (1.6.10).

Example 4. Rooks on chessboards

For n fixed, a *chessboard* C is a subset of $[n] \times [n]$. We are given C, and we define a sequence $\{r_k\}$ as follows: r_k is the number of ways we can place k nonattacking (i.e., no two in the same row or column) rooks on C.

Next, let σ be a permutation of n letters. For each j we let e_j denote the number of permutations that 'meet the chessboard C in exactly j squares,' i.e., if the event $(i, \sigma(i)) \in C$ occurs for exactly j values of i, $1 \leq i \leq n$.

The question is, how can we find the e_j's in terms of the r_k's?

Let the objects Ω be the $n!$ permutations of $[n]$. There will be a property $P(s)$ corresponding to each square $s \in C$. A permutation σ has property $P(s)$ if σ meets the mini-chessboard that consists of the single cell s.

Let S be a set of properties, i.e., of cells in C, and consider the sum $N_k = \sum_{|S|=k} N(\supseteq S)$. Each arrangement of k nonattacking rooks on C contributes $(n-k)!$ to this sum. Indeed, when the set S corresponds to the cells on which those rooks can be placed, then we are looking at k of the n values of a permutation that hits C in at least k squares. The permutation can be completed, in the remaining $n - k$ rows, in $(n - k)!$ ways.

Hence $N_k = r_k(n - k)!$, for each k, $0 \leq k \leq n$. Therefore

$$N(x) = \sum_k (n-k)!r_k x^k,$$

and immediately we find that the number of n-permutations that hit C in exactly j cells is

$$[x^j]\sum_k (n-k)!r_k(x-1)^k. \qquad (4.2.17)$$

Example 5. A problem on subsets.

This example is more cute than profound, but we will at least finish with a combinatorial proof of an interesting identity, as well as illustrating the generating function aspect of the sieve method.

For a fixed positive n, take as our set Ω of objects the $\binom{2n}{n}$ ways of choosing an n-subset of $[2n]$. For the set P of properties we take the following list of n (not $2n$) properties: an n-subset Q has property i if $i \notin Q$, for each $i = 1, 2, \ldots, n$ (note that we are working with only the first half of the possible elements of S).

If S is a set of properties (i.e., is a set of letters chosen from $[n]$), then the number of 'objects' Q that have at least that set of properties (i.e., are missing at least all of the $i \in S$) is clearly

$$N(\supseteq S) = \binom{2n - |S|}{n}.$$

Hence

$$N_r = \sum_{|S|=r} N(\supseteq S) = \binom{n}{r}\binom{2n - r}{n}.$$

If we substitute these N's into the sieve (4.2.5) we find that

$$\sum_j e_j t^j = \sum_r \binom{n}{r}\binom{2n - r}{n}(t - 1)^r. \qquad (4.2.18)$$

This formula tells us the number e_j of objects that have *exactly* j properties, for each j.

But we didn't need to be told that!

An object that has exactly j of these properties is a subset Q of $[2n]$ that is missing exactly j of the elements $1, 2, \ldots, n$. Obviously there are just $\binom{n}{j}^2$ such subsets Q, because we can choose the j elements that they are missing in $\binom{n}{j}$ ways, and we can then choose the other j elements that are needed to fill the subset from $n + 1, \ldots, 2n$ in $\binom{n}{j}$ ways also.

Thus, with no assistance from the sieve method, we already knew that $e_j = \binom{n}{j}^2$, for all j. Hence, according to (4.2.18), it must be true that

$$\sum_j \binom{n}{j}^2 t^j = \sum_r \binom{n}{r}\binom{2n - r}{n}(t - 1)^r. \qquad (4.2.19)$$

We therefore have an odd kind of a combinatorial proof of the identity (4.2.19). The reader should suspect that something of this sort is going on whenever an identity involves an expansion around the origin on one side, and an expansion around $t = 1$ on the other side.

4.3 The 'Snake Oil' method for easier combinatorial identities

Combinatorial mathematics is full of dazzling identities. Legions of them involving binomial coefficients alone fill text- and reference books (see below for some references). It is a fine skill for a working discrete mathematician to have if he/she is able to evaluate or simplify complicated looking sums that involve combinatorial numbers, because they have a way of turning up in connection with problems in graphs, algorithms, enumeration, etc. (they're fun, too!).

In the past, one had to have built up a certain arsenal of special devices, the more the better, in order to be able to trot out the correct one for the correct occasion. Recently, however, a good deal of quite dramatic systematization has taken place, and there are unified methods for handling vast sub-legions of the legions referred to above.

In this section we are going to do two things. First we will give a single method (the *Snake Oil** method) that uses generating functions to deal with the evaluation of combinatorial sums. That one method is capable of handling a great variety of sums involving binomial coefficients, but there's nothing special about binomial coefficients in this respect. The method also works beautifully, within its limitations, on sums involving other combinatorial numbers. The philosophy is roughly this: don't try to evaluate the sum that you're looking at. Instead, find the generating function for the whole parameterized family of them, then read off the coefficients.

Second, we will confess that Snake Oil doesn't cure them all. Some combinatorial sums are really hard. Many of the very hardest binomial coefficient sums can now be proved by computers using the method of rational functions, which we will discuss next. Not only that, but use of the computer has resulted in some new proofs of classical identities. The hallmarks of these proofs are that (a) they are very short compared to the previously known proofs, (b) they seem extremely unmotivated to the reader, but (c) nothing is left out, and they really are proofs. The computerized proof techniques rely on a very simple-looking observation, which we will describe and illustrate.

Therefore, in this section you can expect to see one unified method that works on a lot of relatively easy sums, and one other unified method that works on many more kinds of binomial coefficient sums, including some fiendishly difficult ones.

First let's talk about the Snake Oil method.

The basic idea is what I might call the *external* approach to identities

---

* The Random House Dictionary of the English Language defines 'snake oil' as a purported cure for everything, and gives the example *The governor promised to lower taxes, but it was the same old snake oil.* The date of the expression is given as '1925-30, Amer.'

rather than the usual *internal* method.

To explain the difference between these two points of view, suppose we want to prove some identity that involves binomial coefficients. Typically such a thing would assert that some fairly intimidating-looking sum is in fact equal to such-and-such a simple function of n.

One approach that is now customary, thanks to the skillful exposition and deft handling by Knuth in [K1], and by Graham, Knuth and Patashnik in [GKP], consists primarily of looking inside the summation sign ('internally'), and using binomial coefficient identities or other manipulations of indices *inside* the summations to bring the sum to manageable form.

The method that we are about to discuss is complementary to the internal approach. In the *external*, or generatingfunctionological, approach that we are selling here, one begins by giving a quick glance at the expression that is inside the summation sign, just long enough to spot the 'free variables,' i.e., what it is that the sum depends on after the dummy variables have been summed over. Suppose that such a free variable is called n.

Then instead of trying to grapple with the sum, just sweep it all under the rug, as follows:

The Snake Oil Method for Doing Combinatorial Sums

(a) Identify the free variable, say n, that the sum depends on. Give a name to the sum that you are working on; call it $f(n)$.

(b) Let $F(x)$ be the opsgf whose $[x^n]$ is $f(n)$, the sum that you'd love to evaluate.

(c) Multiply the sum by x^n, and sum on n. Your generating function is now expressed as a double sum over n, and over whatever variable was first used as a dummy summation variable.

(d) Interchange the order of the two summations that you are now looking at, and perform the inner one in simple closed form. For this purpose it will be helpful to have a catalogue of series whose sums are known, such as the list in section 2.5 of this book.

(e) Try to identify the coefficients of the generating function of the answer, because those coefficients are what you want to find.

If that seems complicated, just wait till you see the next seven examples. By then it will seem quite routine.

The success of the method depends on favorable outcomes of steps (d) and (e). What is surprising is the high success rate. It also has the 'advantage' of requiring hardly any thought at all; when it works, you know it, and when it doesn't, that's obvious too.

We will adhere strictly to the customary conventions about binomial coefficients and the ranges of summation variables. These are: first that the binomial coefficient $\binom{x}{m}$ vanishes if $m < 0$ or if x is a nonnegative integer that is smaller than m. Second, a summation variable whose range is not otherwise explicitly restricted is understood to be summed from $-\infty$ to ∞. Thus we have, for integer $n \geq 0$,

$$\sum_k \binom{n}{k} = 2^n,$$

in the sense that the sum ranges over all positive and negative and 0 values of k, the summand vanishes unless $0 \leq k \leq n$, and the sum has the value advertised. These conventions will save endless fussing over changing limits of summation when the dummy variables of summation get changed. For example, we find that

$$\sum_k \binom{n}{r+k} x^k = x^{-r} \sum_k \binom{n}{r+k} x^{r+k} = x^{-r} \sum_s \binom{n}{s} x^s = x^{-r}(1+x)^n,$$

for nonnegative integer n and integer r, without ever even thinking about the ranges of the summation variables.

The series evaluations that are most helpful in the examples that follow are, first and foremost,

$$\sum_{r \geq 0} \binom{r}{k} x^r = \frac{x^k}{(1-x)^{k+1}} \qquad (k \geq 0), \qquad (4.3.1)$$

which is basically a rewrite of (2.5.7). Also useful are the binomial theorem

$$\sum_r \binom{n}{r} x^r = (1+x)^n \qquad (4.3.2)$$

and (2.5.11), which we repeat here for easy reference:

$$\sum_n \frac{1}{n+1} \binom{2n}{n} x^n = \frac{1}{2x}(1 - \sqrt{1-4x}). \qquad (4.3.3)$$

Example 1. Openers

Consider the sum

$$\sum_{k \geq 0} \binom{k}{n-k} \qquad (n = 0, 1, 2, \ldots).$$

The free variable is n, so let's call the sum $f(n)$. Write it out like this:

$$f(n) = \sum_{k \geq 0} \binom{k}{n-k}.$$

OK, now multiply both sides by x^n and sum over n. You have now arrived at step (c) of the general method, and you are looking at

$$F(x) = \sum_n x^n \sum_{k \geq 0} \binom{k}{n-k}.$$

Ready for step (d)? Interchange the sums, to get

$$F(x) = \sum_{k \geq 0} \sum_n \binom{k}{n-k} x^n.$$

We would like to 'do' the inner sum, the one over n. The trick is to get the exponent of x to be exactly the same as the index that appears in the binomial coefficient. In this example the exponent of x is n, and n is involved in the downstairs part of the binomial coefficient in the form $n-k$. To make those the same, the correct medicine is to multiply inside the sum by x^{-k} and outside the inner sum by x^k, to compensate. The result is

$$F(x) = \sum_{k \geq 0} x^k \sum_n \binom{k}{n-k} x^{n-k}.$$

Now the exponent of x is the same as what appears downstairs in the binomial coefficient. Hence take $r = n - k$ as the new dummy variable of summation in the inner sum. We find then

$$F(x) = \sum_{k \geq 0} x^k \sum_r \binom{k}{r} x^r.$$

We recognize the inner sum immediately, as $(1+x)^k$. Hence

$$F(x) = \sum_{k \geq 0} x^k (1+x)^k = \sum_{k \geq 0} (x + x^2)^k = \frac{1}{1 - x - x^2}.$$

The generating function on the right is an old friend; it generates the Fibonacci numbers (see Example 1.3 of chapter 1). Hence $f(n) = F_{n+1}$, and we have discovered that

$$\sum_{k \geq 0} \binom{k}{n-k} = F_{n+1} \qquad (n = 0, 1, 2, \ldots).$$

■

Example 2. Another one

Consider the sum

$$\sum_k \binom{n+k}{m+2k}\binom{2k}{k}\frac{(-1)^k}{k+1} \qquad (m,n \geq 0). \qquad (4.3.4)$$

Can it be that the same method will do this sum, without any further infusion of ingenuity? Indeed; just pour enough Snake Oil on it and it will be cured. Let $f(n)$ denote the sum in question, and let $F(x)$ be its opsgf. Dive in immediately by multiplying by x^n and summing over $n \geq 0$, to get

$$\begin{aligned}
F(x) &= \sum_{n\geq 0} x^n \sum_k \binom{n+k}{m+2k}\binom{2k}{k}\frac{(-1)^k}{k+1} \\
&= \sum_k \binom{2k}{k}\frac{(-1)^k}{k+1}x^{-k}\sum_{n\geq 0}\binom{n+k}{m+2k}x^{n+k} \\
&= \sum_k \binom{2k}{k}\frac{(-1)^k}{k+1}x^{-k}\sum_{r\geq k}\binom{r}{m+2k}x^{r} \\
&= \sum_k \binom{2k}{k}\frac{(-1)^k}{k+1}x^{-k}\frac{x^{m+2k}}{(1-x)^{m+2k+1}} \qquad \text{(by (4.3.1))} \\
&= \frac{x^m}{(1-x)^{m+1}}\sum_k \binom{2k}{k}\frac{1}{k+1}\left\{\frac{-x}{(1-x)^2}\right\}^k \\
&= \frac{-x^{m-1}}{2(1-x)^{m-1}}\left\{1 - \sqrt{1 + \frac{4x}{(1-x)^2}}\right\} \\
&= \frac{-x^{m-1}}{2(1-x)^{m-1}}\left\{1 - \frac{1+x}{1-x}\right\} \\
&= \frac{x^m}{(1-x)^m}.
\end{aligned}$$

The original sum is now unmasked: it is the coefficient of x^n in the last member above. But that is $\binom{n-1}{m-1}$, by (4.3.1) again, and we have our answer. See exercise 16 for a generalization of this sum.

If the train of manipulations seemed long, consider that at least it's always the *same* train of manipulations, whenever the method is used, and also that with some effort a computer could be trained to do it! ∎

Example 3. A discovery

Is it possible to write the sum

$$f_n = \sum_{k\leq \frac{n}{2}} (-1)^k \binom{n-k}{k} y^{n-2k} \qquad (n \geq 0) \qquad (4.3.5)$$

in a simpler closed form?

This example shows the whole machine at work again, along with a few new wrinkles. The first step is to let $F \xleftarrow{ops} \{f_n\}$, and try to find the generating function F instead of the sequence $\{f_n\}$.

To do that we multiply (4.3.5) on both sides by x^n and sum over $n \geq 0$ to obtain

$$F(x) = \sum_{n \geq 0} x^n \sum_{k \leq \frac{n}{2}} (-1)^k \binom{n-k}{k} y^{n-2k}.$$

The next step is invariably to interchange the summations and hope. To try to make the innermost summation as clean looking as possible, be sure to take to the outer sum any factors that depend only on k. This yields

$$F(x) = \sum_{k} (-1)^k y^{-2k} \sum_{n \geq 2k} \binom{n-k}{k} x^n y^n.$$

Now focus on (4.3.1), and try to make the inner sum look like that. If in our inner sum the powers of x and y were $x^{n-k} y^{n-k}$, then those exponents would match exactly the upper story of the binomial coefficient $\binom{n-k}{k}$, and so after a change of dummy variable of summation we would be looking exactly at the left side of (4.3.1).

Hence we next multiply inside the inner sum by $x^{-k} y^{-k}$, and outside the inner sum by $x^k y^k$. Now we have

$$F(x) = \sum_{k} (-1)^k y^{-2k} x^k y^k \sum_{n \geq 2k} \binom{n-k}{k} x^{n-k} y^{n-k}$$

$$= \sum_{k} (-1)^k x^k y^{-k} \sum_{a \geq k} \binom{a}{k} (xy)^a$$

$$= \sum_{k \geq 0} (-1)^k x^k y^{-k} \frac{(xy)^k}{(1-xy)^{k+1}} \qquad \text{(by (4.3.1))}$$

$$= \frac{1}{1-xy} \sum_{k \geq 0} \left\{ \frac{-x^2}{1-xy} \right\}^k$$

$$= \frac{1}{1-xy} \frac{1}{1 + \frac{x^2}{1-xy}}$$

$$= \frac{1}{1-xy+x^2}.$$

(4.3.6)

(Question: Why, after the third equals sign above, did the range of k get restricted to '$k \geq 0$'?)

We now expand (4.3.6) in partial fractions to obtain a closed form for the sum (4.3.5). This gives

$$F(x) = \frac{1}{(1 - xx_+)(1 - xx_-)}$$

$$= \frac{x_+}{(x_+ - x_-)(1 - xx_+)} - \frac{x_-}{(x_+ - x_-)(1 - xx_-)},$$

where

$$x_\pm = \frac{y \pm \sqrt{y^2 - 4}}{2}.$$

Hence, for $n \geq 0$ the coefficient of x^n is

$$f_n = \frac{1}{\sqrt{y^2 - 4}} \left\{ \left(\frac{y + \sqrt{y^2 - 4}}{2} \right)^{n+1} - \left(\frac{y - \sqrt{y^2 - 4}}{2} \right)^{n+1} \right\}.$$

We now have our answer, but just for a demonstration of the effectiveness of cleanup operations, let's invest a little more time in making the answer look as neat as possible. Because of the ubiquitous appearance of $\sqrt{y^2 - 4}$ in the answer, we replace y formally by $x + (1/x)$. Then

$$\sqrt{y^2 - 4} = x - \frac{1}{x},$$

and our formula becomes

$$\sum_{k \leq \frac{n}{2}} (-1)^k \binom{n - k}{k} (x^2 + 1)^{n-2k} x^{2k} = \frac{x^{2n+2} - 1}{x^2 - 1} \qquad (n \geq 0).$$

Finally we write $t = x^2$ to obtain the pretty evaluation

$$\sum_{k \leq \frac{n}{2}} (-1)^k \binom{n - k}{k} (t + 1)^{n-2k} t^k = \frac{1 - t^{n+1}}{1 - t} \qquad (n \geq 0). \qquad (4.3.7)$$

For instance, the value $t = 1$ gives

$$\sum_{k \leq \frac{n}{2}} (-1)^k \binom{n - k}{k} 2^{n-2k} = n + 1 \qquad (n \geq 0). \qquad (4.3.8)$$

As a final touch, we can read off the coefficient of t^m in (4.3.7) to discover the interesting fact that

$$\sum_{k \leq \frac{n}{2}} (-1)^k \binom{n - k}{k} \binom{n - 2k}{m - k} = \begin{cases} 1, & \text{if } 0 \leq m \leq n; \\ 0, & \text{otherwise.} \end{cases} \qquad (4.3.9)$$

Try this identity with $n = 2$ and watch what happens. ■

Here is another example of the same technique.

Example 4.

Evaluate the sums

$$f_n = \sum_k \binom{n+k}{2k} 2^{n-k} \qquad (n \geq 0). \qquad (4.3.10)$$

Without stopping to think, let F be the opsgf of the sequence, multiply both sides of (4.3.10) by x^n, sum over $n \geq 0$, and interchange the two sums on the right. This produces

$$F = \sum_k 2^{-k} \sum_{n \geq 0} \binom{n+k}{2k} 2^n x^n$$

$$= \sum_k 2^{-k} (2x)^{-k} \sum_{n \geq 0} \binom{n+k}{2k} (2x)^{n+k}$$

$$= \sum_{k \geq 0} 2^{-k} (2x)^{-k} \frac{(2x)^{2k}}{(1-2x)^{2k+1}} \qquad \text{(by (4.3.1))}$$

$$= \frac{1}{1-2x} \sum_{k \geq 0} \left\{ \frac{x}{(1-2x)^2} \right\}^k$$

$$= \frac{1}{1-2x} \frac{1}{1 - \frac{x}{(1-2x)^2}}$$

$$= \frac{1-2x}{(1-4x)(1-x)}$$

$$= \frac{2}{3(1-4x)} + \frac{1}{3(1-x)}.$$

It is now a triviality to read off the coefficient of x^n on both sides and discover the answer:

$$\sum_k \binom{n+k}{2k} 2^{n-k} = \frac{2^{2n+1}+1}{3} \qquad (n \geq 0). \qquad (4.3.11)$$

■

Example 5.

Our next example will be of a sum that we won't succeed in evaluating in a neat, closed form. However, the generating function that we obtain will be rather tidy, and that is about the most that can be expected from this family of sums.

The sum is

$$f_n(y) = \sum_k \binom{n}{k}\binom{2k}{k} y^k \qquad (n \geq 0). \tag{4.3.12}$$

Follow the usual prescription. Define $F(x,y) = \sum_{n\geq 0} f_n(y)x^n$. To find F, multiply (4.3.12) by x^n, sum over $n \geq 0$ and interchange the inner and outer sums, to obtain

$$F(x,y) = \sum_k \binom{2k}{k} y^k \sum_{n\geq 0} \binom{n}{k} x^n$$

$$= \sum_k \binom{2k}{k} y^k \frac{x^k}{(1-x)^{k+1}} \tag{4.3.13}$$

$$= \frac{1}{1-x} \sum_k \binom{2k}{k}\left(\frac{xy}{1-x}\right)^k.$$

Now since

$$\sum_k \binom{2k}{k} z^k = \frac{1}{\sqrt{1-4z}}, \tag{4.3.14}$$

by (2.5.12), we obtain

$$F(x,y) = \frac{1}{(1-x)\sqrt{1-\frac{4xy}{1-x}}}$$

$$= \frac{1}{\sqrt{(1-x)(1-x(1+4y))}}. \tag{4.3.15}$$

For general values of y, that's about all we can expect. There are two special values of y for which we can go further. If $y = -1/4$, we find that

$$\sum_k \binom{2k}{k}\binom{n}{k}(-\frac{1}{4})^k = 2^{-2n}\binom{2n}{n} \qquad (n \geq 0). \tag{4.3.16}$$

If $y = -1/2$, then

$$F(x,-1/2) = 1/\sqrt{1-x^2}$$

$$= \sum_m \binom{2m}{m}(x/2)^{2m} \qquad \text{(by (2.5.11))}.$$

Hence we have Reed Dawson's identity

$$\sum_k \binom{2k}{k}\binom{n}{k}(-1)^k 2^{-k} = \begin{cases} \binom{n}{n/2} 2^{-n} & \text{if } n \geq 0 \text{ is even,} \\ 0 & \text{if } n \geq 0 \text{ is odd,} \end{cases} \tag{4.3.17}$$

and Snake Oil triumphs again. ∎

Example 6.

Suppose we have two complicated sums and we want to show that they're the same. Then the generating function method, if it works, should be very easy to carry out. Indeed, one might just find the generating functions of each of the two sums independently and observe that they are the same.

Suppose we want to prove that

$$\sum_k \binom{m}{k}\binom{n+k}{m} = \sum_k \binom{m}{k}\binom{n}{k} 2^k \qquad (m, n \geq 0)$$

without evaluating either of the two sums.

Multiply on the left by x^n, sum on $n \geq 0$ and interchange the summations, to arrive at

$$\sum_k \binom{m}{k} x^{-k} \sum_{n\geq 0} \binom{n+k}{m} x^{n+k} = \sum_k \binom{m}{k} x^{-k} \frac{x^m}{(1-x)^{m+1}}$$

$$= \frac{x^m}{(1-x)^{m+1}} \left(1 + \frac{1}{x}\right)^m$$

$$= \frac{(1+x)^m}{(1-x)^{m+1}}.$$

If we multiply on the right by x^n, etc., we find

$$\sum_k \binom{m}{k} 2^k \sum_{n\geq 0} \binom{n}{k} x^n = \frac{1}{(1-x)} \sum_k \binom{m}{k} \left(\frac{2x}{(1-x)}\right)^k$$

$$= \frac{1}{(1-x)} \left(1 + \frac{2x}{1-x}\right)^m$$

$$= \frac{(1+x)^m}{(1-x)^{m+1}}.$$

Hence the two sums are equal, even if we don't know what they are! ∎

Example 7.

There are, in combinatorics, a number of *inversion formulas*, and generating functions give an easy way to prove many of those. An inversion formula in general is a relationship that expresses one sequence in terms of another, along with the inverse relation, which recovers the original sequence from the constructed one.

We have already seen a couple of famous examples of these. One is the Möbius inversion formula, which is the pair (2.6.11), (2.6.12). Another is the pair (4.2.3), (4.2.7) that occurred in the sieve method. We repeat that pair here, for ready reference. It states that if we compute a sequence $\{N_r\}$ from a sequence $\{e_r\}$ by the relations

$$N_r = \sum_{t \geq 0} \binom{t}{r} e_t \qquad (r = 0, 1, 2, \ldots), \tag{4.3.18}$$

then we can recover the original sequence ('invert') by means of

$$e_t = \sum_j (-1)^{j-t} \binom{j}{t} N_t \qquad (t \geq 0).$$

To give just one more example of such a pair of formulas, consider the relation

$$a_r = \sum_s \binom{r}{s} b_s \qquad (r \geq 0), \tag{4.3.19}$$

which differs from the previous pair in that the summation is over the lower index in the binomial coefficient. How can we find the relations that are inverse to (4.3.18)? That is, how can we solve for the b's in terms of the a's?

The answer is that we convert the relation (4.3.18) between two sequences into a relation between their exponential generating functions, which we then invert. By (2.3.3) we have $A(x) = e^x B(x)$, where A and B are the egf's. Hence $B(x) = e^{-x} A(x)$, and therefore

$$b_n = \sum_m \binom{n}{m} (-1)^{n-m} a_m \qquad (n \geq 0). \tag{4.3.20}$$

■

An inversion formula of a somewhat deeper kind appears in (5.1.5), (5.1.6).

Example 8. Snake Oil vs. hypergeometric functions.

Many combinatorial identities are special cases of identities in the theory of hypergeometric series (we'll explain that remark, briefly, in a moment). However, the Snake Oil method can cheerfully deal with all sorts of identities that are not basically about hypergeometric functions. So the approaches are complementary.

A hypergeometric series is a series

$$\sum_k T_k$$

in which the *ratio* of every two consecutive terms is a rational function of the summation variable k. That means that

$$\frac{T_{k+1}}{T_k} = \frac{P(k)}{Q(k)},$$

where P and Q are polynomials, and it takes in a lot of territory. Many binomial coefficient identities, including all of the examples in this chapter so far, are of this type. There are some general tools for dealing with such sums, and these are very important considering how frequently they occur in practice. For a discussion of some of these tools, see, for example, the article by Roy [Ro].

In this example we want to emphasize that the scope of the Snake Oil method includes a lot of sums that are not hypergeometric. Consider, for instance, the following sum-

$$f(n) = \sum_k \begin{bmatrix} n \\ k \end{bmatrix} B_k,$$

where the $\begin{bmatrix} \ \end{bmatrix}$'s are the Stirling numbers of the first kind, and the B's are the Bernoulli numbers.

Now one thing, at least, is clear from looking at this sum: it is not hypergeometric. The ratio of two consecutive terms is certainly not a rational function of k. The Snake Oil method is, however, unfazed by this turn of events. If you follow the method exactly as before, you could define $F(x)$ to be the egf of the sequence $\{f(n)\}$, multiply by $x^n/n!$, sum on n, interchange the indices, etc., and obtain

$$F(x) = \sum_n \frac{f(n)x^n}{n!}$$

$$= \sum_n \frac{x^n}{n!} \sum_k \begin{bmatrix} n \\ k \end{bmatrix} B_k$$

$$= \sum_k B_k \sum_n \begin{bmatrix} n \\ k \end{bmatrix} \frac{x^n}{n!}$$

$$= \sum_k B_k \left\{ \frac{1}{k!} \left(\log \frac{1}{1-x} \right)^k \right\} \qquad \text{(by (3.5.3))}$$

$$= \sum_k \frac{B_k}{k!} u^k \qquad (u = \log \frac{1}{1-x})$$

$$= \frac{u}{e^u - 1} \qquad \text{(by (2.5.8))}$$

$$= \frac{1-x}{x} \log \frac{1}{1-x}.$$

If we now read off the coefficient of $x^n/n!$ on both sides, we find that the unknown sum is

$$\sum_k \begin{bmatrix} n \\ k \end{bmatrix} B_k = -\frac{(n-1)!}{n+1} \qquad (n \geq 1). \tag{4.3.21}$$

Example 9. The scope of the Snake Oil method

The success of the Snake Oil method depends upon being given a sum to evaluate in which there is a free variable that appears in only one place. Then, after interchanging the order of the summations, one finds one of the basic power series (4.3.1) or (4.3.2) to sum.

At the risk of diminishing the charm of the method somewhat by adding gimmicks to it, one must remark that in many important cases this limitation on the scope is easy to overcome. This is because it frequently happens that when an identity is presented that has a free variable repeated several times, that identity turns out to be a special case of a more general identity in which each of the repeated appearances of the free variable is replaced by a *different* free variable. Before abandoning the method on some given problem, this possibility should be explored.

Consider the identity

$$\sum_i \binom{n}{i}\binom{2n}{n-i} = \binom{3n}{n}.$$

At first glance the possibilities for successful Snake Oil therapy seem dim because of the multiple appearances of n in the summand. However, if we generalize the identity by splitting the appearances of n into different free variables, we might be led to consider the sum

$$\sum_i \binom{n}{i}\binom{m}{r-i},$$

which is readily evaluated by the Snake Oil method. It is characteristic of the subject of identities that it is usually harder to prove special cases than general theorems. Multiple appearances of a free variable are often a hint that one should try to find a suitable generalization.

4.4 WZ pairs prove harder identities

Computers can now *find proofs of* combinatorial identities, including most of the identities that we did by the Snake Oil method in the previous section, as well as many, many more. In this section we will say how that is done. Although *finding* proofs this way requires more work than a human would care to do, the result, after the computer is finished, is a neat and compact proof that a human can often easily check, and can always check with the assistance of one of the numerous symbolic manipulation programs that are now available on personal computers. Hence there is no need for blind trust in the computer. One can ask it to find a proof of an identity, and one can readily check that the proof is correct.

These developments are quite recent, and they will surely change our attitudes towards, for instance, binomial coefficient identities. Instead of regarding each one as a challenge to our ingenuity, we can instead ask our computer to find a proof. It will not always succeed, but in the vast majority of cases it will. In fact, even more powerful methods are now becoming available, which promise a 100% success rate in certain classes of identities.

This doesn't mean that it was a waste of time to have learned the Snake Oil method. There were identities that Snake Oil handled that the method of [WZ] (which we're about to discuss) cannot deal with, like the fact that $\sum_{k \geq 0} \binom{k}{n-k} = F_n$ for $n \geq 0$, which was the first example in the previous section. Another one that offers no hope to the WZ method is (4.3.21), which involves Stirling and Bernoulli numbers. Also, to use the Snake Oil method, one doesn't need to know the right hand side of the identity in advance; the method will find it. The method that we are about to describe will prove a given identity, but it won't discover the identity for itself.

With those disclaimers, however, it is fair to say that the method is quite versatile, and seems able to handle in a unified way some of the knottiest identities that have ever been discovered.

It stems from some totally obvious facts, concatenated in a slightly un-obvious way. Suppose we want to prove that an identity

$$\sum_k U(n,k) = rhs(n) \qquad (n = 0, 1, 2, \ldots)$$

is true. The first thing we do is to divide by the right hand side to get the *standard form*

$$\sum_k F(n,k) = 1 \qquad (n = 0, 1, 2, \ldots). \tag{4.4.1}$$

So, in standard form, we are trying to prove that a certain sum is independent of n, for $n \geq 0$.

To do that, rewrite (4.4.1) with n replaced by $n+1$ and then subtract (4.4.1), to get

$$\sum_k \{F(n+1,k) - F(n,k)\} = 0 \qquad (n = 0, 1, \ldots) \tag{4.4.2}$$

Wouldn't it be helpful if there were a nice function $G(n,k)$ such that

$$F(n+1,k) - F(n,k) = G(n,k+1) - G(n,k), \tag{4.4.3}$$

for then the sum (4.4.2) would telescope? In detail, that would mean that

$$\sum_{k=-L}^{k=K} \{F(n+1,k) - F(n,k)\} = \sum_{k=-L}^{k=K} \{G(n,k+1) - G(n,k)\}$$

$$= G(n, K+1) - G(n, -L).$$

Well, as long as we're wishing, why not wish for $G(n, \pm\infty) = 0$ too, for then, by letting $K, L \to \infty$ we would find that (4.4.2) is indeed true! This line of reasoning leads quickly to the following:

Theorem 4.4.1. *(Wilf, Zeilberger [WZ]) Let (F, G) satisfy (4.4.3), and suppose*

$$\lim_{k \to \pm\infty} G(n, k) = 0 \qquad (n = 0, 1, 2, \dots). \tag{4.4.4}$$

Then the identity

$$\sum_k F(n, k) = const. \qquad (n = 0, 1, 2, \dots)$$

holds. ∎

Example 1.

Suppose we want to prove the identity

$$\sum_k \binom{n}{k} = 2^n \qquad (n \geq 0). \tag{4.4.5}$$

If we divide by the right hand side we find that the function $F(n, k)$ of (4.4.1) is

$$F(n, k) = \binom{n}{k} / 2^n \qquad (n \geq 0). \tag{4.4.6}$$

Now we need to find the mate $G(n, k)$ of this F. Well, it is

$$G(n, k) = -\frac{\binom{n}{k-1}}{2^{n+1}}. \tag{4.4.7}$$

That raises two questions. Does this G really work, and where did it come from?

Let's take the easy one first. To check that it works we need to check first that (4.4.3) holds, which in this case says that

$$2^{-n-1}\binom{n+1}{k} - 2^{-n}\binom{n}{k} = -2^{-n-1}\binom{n}{k} + 2^{-n-1}\binom{n}{k-1}.$$

After a few moments of work we can satisfy ourselves that this is true. Further, the boundary conditions (4.4.4) are easy to verify, and we are all finished. ∎

Definition. *We say that an identity (4.4.1) is certified by a pair (F, G) ('WZ pair') if the conditions (4.4.3), (4.4.4) hold.*

Hence, the simple identity (4.4.5) is certified by the pair (F, G) of (4.4.6), (4.4.7).

We can cut down still further on the complexity of the apparatus, as follows. It will turn out in the class of identities that we are discussing here that the mate G will always be of the form $G(n, k) = R(n, k)F(n, k - 1)$, where $R(n, k)$ is a rational function of n and k. Hence, instead of describing the pair (F, G), we need to give only F and R. But F comes directly from the identity that we're trying to prove, just by dividing the summand by the right hand side. So if we have the identity in front of us, then the rational function R is the only extra certification that we need.

The class of identities for which the above simplification is true is the class of those that are of the form (4.4.1) in which the function F has the property that *both*

$$\frac{F(n + 1, k)}{F(n, k)} \quad \text{and} \quad \frac{F(n, k - 1)}{F(n, k)}$$

are rational functions of n and k. This class includes just about every binomial coefficient identity that we have encountered or will encounter.

Let's recapitulate the complete proof procedure for an identity that is certified by a single rational function $R(n, k)$:

(a) Given an identity $\sum_k U(n, k) = rhs(n)$ $(n = 0, 1, 2, \ldots)$, and also given a rational function proof certificate $R(n, k)$.

(b) Define $F(n, k) = U(n, k)/rhs(n)$, for $n \geq 0$ and integer k.

(c) Define $G(n, k) = R(n, k)F(n, k - 1)$.

(d) Check that the conditions (4.4.3) and (4.4.4) are satisfied.

(e) Check that the identity is true when $n = 0$.

(f) The proof of the identity is now complete. ∎

Now here are some more examples of the technique in action. Do take the time to check one or more of these using the full proof technique (a)-(f) given above.

Theorem. $\sum_k (-1)^k \binom{n}{k} \binom{2k}{k} 4^{n-k} = \binom{2n}{n}$ $(n \geq 0)$

Proof: Take $R(n, k) = (2k - 1)/(2n + 1)$. ∎

Let's check that statement, one step at a time, following the proof procedure above. From step (b) we find that

$$F(n, k) = (-1)^k \frac{\binom{n}{k} \binom{2k}{k} 4^{n-k}}{\binom{2n}{n}}.$$

From step (c) we find that

$$G(n,k) = (-1)^{k-1} \frac{2k-1}{2n+1} \binom{n}{k-1} \binom{2k-2}{k-1} \frac{4^{n-k+1}}{\binom{2n}{n}}.$$

Now we know the pair (F, G). In step (d) we must first check that the condition $F(n+1, k) - F(n, k) = G(n, k+1) - G(n, k)$ is satisfied. At this point it would be very helpful to have a symbolic manipulation program available, for then one would simply type in the functions F and G, and ask it to verify that the condition holds. Otherwise, it's a rather dull pencil and paper computation of five minutes' length, and we will omit it.

The second part of step (d) is the check of the boundary condition

$$\lim_{k \to \pm\infty} G(n, k) = 0 \qquad (n = 0, 1, 2, \ldots).$$

That, however, is a triviality, because in this case not only are the limits 0, but the function $G(n, k)$ is 0 for every single value of $k > n + 1$ and for all values of $k < 1$.

In step (e) we quickly check and find that 1=1, and the proof is complete. ∎

Theorem. $\sum_k (-1)^k \binom{n}{k} / \binom{k+a}{k} = a/(n+a)$ $\qquad (n \geq 0)$

Proof: Take $R(n, k) = k/(n+a)$. ∎

Theorem. $\sum_k (-1)^{n-k} \binom{2n}{k}^2 = \binom{2n}{n}$ $\qquad (n \geq 0)$

Proof: Take $R(n, k) = -(10n^2 - 6kn + 17n + k^2 - 5k + 7)/(2(2n - k + 2)^2)$. ∎

After the last three examples the reader will probably be wondering how to *find* the $R(n, k)$'s, instead of just checking that an $R(n, k)$ snatched out of the blue sky seems to work. So we are going to tell that story too, because it's a very important one, not just for these purposes but for symbolic manipulation in general.

The problem is this: the function $F(n, k)$ is known, and we want to find $G(n, k)$ so that the identity (4.4.3) is true. Since F is known, so is $F(n+1, k) - F(n, k)$, so we might as well call it $f(n, k)$. Next, observe that at this moment the index n is a silent partner. That is, we are looking for G so that $f(n, k) = G(n, k+1) - G(n, k)$, and we see that k is an active index but n is just a parameter, since n has the same value throughout. So we might as well suppress the appearance of n altogether, and state the problem this way: if f_k is a given function of k (and other parameters), how can we find g_k so that $f_k = g_{k+1} - g_k$ for all integers k?

We still don't quite have the right question, but we're getting there. The question as asked is a triviality. There is always such a g_k and it is just $\sum_{j<k} f_j$. Take the function $f_k = k$, for instance. Then we can take g_k to be $\sum_{0 \le j < k} j$.

To ask the right question we have to add the condition that the sum that represents g_k can be done *in closed form*. This whole problem is about closed forms. What is closed form? Well roughly it means that the answer should be pleasant to look at and have no summation signs left in it. That idea is too nebulous to work with, so we will use one way of making it precise that has proved to be productive.

Definition. *A function f_k of the integer k is a hypergeometric term if f_{k+1}/f_k is a rational function of k.*

Thus $k!$ is a hypergeometric term. So is $(3k+2)!/(5k-6)!$, and so is

$$(-1)^k 4^{n-k} \frac{\binom{2k}{k}}{\binom{n+k}{k}}.$$

The function k^k is not a hypergeometric term, nor is $e^{\sqrt{k}}$. The function

$$f_k = \sum_{0 \le j \le k} \binom{n}{j}$$

is not obviously a hypergeometric term, nor is it obviously not a hypergeometric term. The expression serves to define the function but does not immediately reveal the nature of the beast.

Now we're ready for the right question.

Let f_k be defined for integer k and be a hypergeometric term. Does there exist a hypergeometric term g_k such that $f_k = g_{k+1} - g_k$ for all integers k?

An algorithm that is due to R. W. Gosper, Jr. [Gos] gives a complete algorithmic answer to this question. That is to say, if we input f to Gosper's algorithm it will then either return a function g with the desired properties, or it will return a guarantee that no such function exists. It will *not* ever return a statement 'I don't know.' We will not describe Gosper's algorithm here because that would take us rather far afield from generating functions. However, the reader is urged to consult either the original reference [Gos] or the lucid explanation in [GKP].

Gosper's algorithm is built in to some of the commercially available symbolic manipulation packages. At this writing (June, 1989) the algorithm is fully implemented in Macsyma, where it can be invoked with the *nusum* command, and it is partially implemented in Mathematica. No doubt it will be more widely available as its usefulness becomes recognized.

Now here are the statements and proofs of two more general and difficult identities, using the [WZ] method of proof by rational function certification.

Theorem 4.4.2. *(The Pfaff-Saalschütz identity)*

$$\sum_k \frac{(a+k)!(b+k)!(c-a-b+n-1-k)!}{(k+1)!(n-k)!(c+k)!} =$$

$$\frac{(a-1)!(b-1)!(c-a-b-1)!(c-a+n)!(c-b+n)!}{(c-a-1)!(c-b-1)!(n+1)!(c+n)!}.$$

Proof: Take

$$R(n,k) = -\frac{(b+k)(a+k)}{(c-b+n+1)(c-a+n+1)}.$$

■

Theorem 4.4.3. *(Dixon's identity)*

$$\sum_k (-1)^k \binom{n+b}{n+k}\binom{n+c}{c+k}\binom{b+c}{b+k} = \frac{(n+b+c)!}{n!b!c!}.$$

Proof: Take $R(n,k) = (c+1-k)(b+1-k)/(2(n+k)(n+b+c+1))$.   ■

4.5 Generating functions and unimodality, convexity, etc.

The binomial coefficients are the prototype of *unimodal* sequences. A sequence is unimodal if its entries rise to a maximum and then decrease. The binomial coefficients $\{\binom{n}{k}\}_{k=0}^n$ do just that. The maximum ('mode') of the binomial coefficient sequence occurs at $k = n/2$ if n is even, and at $k = (n \pm 1)/2$ if n is odd.

In general, a sequence $c_0, c_1, \ldots, c_n$ is unimodal if there exist indices r, s such that

$$c_0 \le c_1 \le c_2 \le \cdots \le c_r = c_{r+1} = \cdots = c_{r+s} \ge c_{r+s+1} \ge \cdots \ge c_n. \quad (4.5.1)$$

Many of the sequences that occur in combinatorics are unimodal. Sometimes it is easy and sometimes it can be very hard to prove that a given sequence is unimodal. Generating functions can help with this kind of a problem, though they are far from a panacæa.

A stronger property than unimodality is *logarithmic concavity*. First recall that a function f on the real line is concave if whenever $x < y$ we have $f((x+y)/2) \ge (f(x) + f(y))/2$. This means that the graph of the function bulges up over every one of its chords.

Similarly, a sequence $c_0, c_1, \ldots, c_n$ of positive numbers is log concave if $\log c_\mu$ is a concave function of μ, which is to say that

$$(\log c_{\mu-1} + \log c_{\mu+1})/2 \le \log c_\mu.$$

If we exponentiate both sides of the above, to eliminate all of the logarithms, we find that the sequence is log concave if

$$c_{\mu-1}c_{\mu+1} \le c_\mu^2 \qquad (\mu = 1, 2, \ldots, n-1). \quad (4.5.2)$$

If, in (4.5.2) we can replace the '$\le$' by '$<$', then we will say that the sequence is *strictly* log concave.

Proposition. *Let $\{c_r\}_0^n$ be a log concave sequence of positive numbers. Then the sequence is unimodal.*

Proof. If the sequence is not unimodal then it has three consecutive members that satisfy $c_{r-1} > c_r < c_{r+1}$ which contradicts the assumed log concavity. ∎

In many cases of interest, generating functions can help to prove log concavity of a sequence, and therefore unimodality too. The source of such results is usually some variant of the following:

Theorem 4.5.2. *Let $p(x) = c_0 + c_1 x + c_2 x^2 + \cdots + c_n x^n$ be a polynomial all of whose zeros are real and negative. Then the coefficient sequence $\{c_r\}_0^n$ is strictly log concave.*

To prove the theorem we need to recall Rolle's theorem of elementary calculus. It holds that if $f(x)$ is continuously differentiable in (a, b), and if $f(a) = f(b)$, then somewhere between a and b the derivative f' must vanish.

If f is a polynomial this can be considerably strengthened. Let u and v be two consecutive distinct zeros of f. Then by Rolle's theorem there is a zero of f' in (u, v). Suppose f is of degree n, has only real zeros, and has exactly r *distinct* real zeros. Then Rolle's theorem accounts for $r - 1$ of the zeros of f', because we find one between each pair of consecutive distinct zeros of f. The remaining $n - r$ zeros of f are copies of the distinct zeros. But if x_0 is a root of f of multiplicity $m > 1$, then $(x - x_0)^m$ is a factor of f, and so $(x - x_0)^{m-1}$ is a factor of f'. Thus x_0 is a zero of multiplicity $m - 1$ of f'. This accounts for the other $n - 1 - (r - 1) = n - r$ zeros of f'. In particular, *the zeros of f' are all real if the zeros of f are.* For maximum utility in our present discussion, we summarize this discussion in the following way:

Lemma 4.5.1. *Let*

$$f(x, y) = c_0 x^n + c_1 x^{n-1} y + \cdots + c_n y^n \tag{4.5.3}$$

be a polynomial all of whose roots x/y are real. Let $g(x, y)$ be the result of differentiating f some number of times with respect to x and y. If g is not identically zero, then all of its zeros are real.

Proof of theorem 4.5.2: Since the zeros of f are all negative, we have

$$f(x) = c_0 + c_1 x + \cdots + c_n x^n = \prod_{j=1}^{n} (x + x_j), \tag{4.5.4}$$

where the x_j's are positive real numbers. Hence none of the c_i's can vanish. Now apply the differential operator $D_x^m D_y^{n-m-2}$ to the polynomial $f(x, y)$ of (4.5.3). Then only three terms survive, viz.:

$$\frac{c_{n-m-2}(m+2)}{n-m-1} x^2 + 2c_{n-m-1} xy + \frac{(n-m)c_{n-m}}{m+1} y^2. \tag{4.5.5}$$

We can put this in a cleaner form by writing $c_j = \binom{n}{j}p_j$, in which case the result (4.5.5) becomes

$$\binom{n}{m+1}(p_{n-m-2}x^2 + 2p_{n-m-1}xy + p_{n-m}y^2).$$

But this quadratic polynomial, according to lemma 4.5.1 above, must have two real roots, and so its discriminant must be nonnegative, i.e.,

$$p_{n-m-1}^2 \geq p_{n-m-2}p_{n-m},$$

and the sequence of p's is log concave. If we substitute back the c's, we find that

$$c_{n-m-1}^2 \geq \frac{(m+2)(n-m)}{(m+1)(n-m-1)}c_{n-m-2}c_{n-m}$$
$$> c_{n-m-2}c_{n-m},$$

and the strict log concavity is established. ∎

Corollary 4.5.1. *The binomial coefficient sequence* $\left\{\binom{n}{k}\right\}_{k=0}^{n}$ *is log concave, and therefore unimodal.*

Proof. The zeros of the generating polynomial $(1+x)^n$ are evidently real and negative. ∎

Corollary 4.5.2. *The sequence of Stirling numbers of the first kind*

$$\{\begin{bmatrix}n\\k\end{bmatrix}\}_{k=1}^{n}$$

is log concave, and therefore unimodal.

Proof. According to (3.5.2), the opsgf of these Stirling numbers is the polynomial

$$\sum_j \begin{bmatrix}n\\j\end{bmatrix}x^{j-1} = (x+1)(x+2)\cdots(x+n-1),$$

whose zeros are clearly real and negative. ∎

Corollary 4.5.3. *The sequence of Stirling numbers of the second kind* $\{\begin{Bmatrix}n\\k\end{Bmatrix}\}_{k=1}^{n}$ *is log concave, and therefore unimodal.*

Proof. We'll have to work just a little harder for this one, because the zeros of the polynomial

$$A_n(x) = \sum_j \begin{Bmatrix}n\\j\end{Bmatrix}x^j$$

are not easy to find. They are, however, real and negative, and here is one way to see that: by (1.6.8) we have the recurrence formula

$$A_n(y) = \{y(1 + D_y)\}A_{n-1}(y) \qquad (n > 0; A_0 = 1),$$

which can be rewritten in the form

$$e^y A_n(y) = y\left(e^y A_{n-1}(y)\right)' \qquad (n > 0; A_0 = 1). \qquad (4.5.6)$$

We claim that for each $n = 0, 1, 2, \ldots$, the function $e^y A_n(y)$ has exactly n zeros, which are real, distinct, and negative except for the one at $y = 0$. This is true for $n = 0$, and if it is true for $0, 1, \ldots, n-1$, then (4.5.6) and Rolle's theorem guarantee that $(e^y A_{n-1}(y))'$ has $n - 2$ negative, distinct zeros, one between each pair of zeros of $A_{n-1}(y)$. After multiplying by y, as in (4.5.6), we have $n - 1$ negative, distinct zeros for $e^y A_n(y)$, but we need to find still one more. But $e^y A_{n-1}(y)$ obviously approaches zero as $y \to -\infty$. Hence its derivative must have one more zero to the left of the leftmost zero of $A_{n-1}(y)$, and we are finished. ∎

The theorem is very strong, but one must not be left with the impression that unimodality or log concavity has something essential to do with reality of the zeros of the generating polynomials. Many sequences are known that are unimodal, and have generating polynomials whose zeros all lie on the unit circle, and are quite uniformly distributed, in angle, around the circle. In such cases our theorem will be of no help.

For example, an *inversion* of a permutation σ of n letters is a pair (i, j) for which $1 \le i < j \le n$, but $\sigma(i) > \sigma(j)$. A permutation may have between 0 and $\binom{n}{2}$ inversions. It is well known that if $b(n, k)$ is the number of permutations of n letters that have exactly k inversions, then

$$\{b(n, k)\}_{k \ge 0} \xrightarrow{ops} (1 + x)(1 + x + x^2) \cdots (1 + x + x^2 + \cdots + x^{n-1}). \quad (4.5.7)$$

The zeros of the generating polynomial are very uniformly sprinkled around the unit circle, so the hypotheses of theorem 4.5.2 are extravagantly violated. Nonetheless, the sequence is unimodal; it rises steadily for $k \le \binom{n}{2}/2$, and falls steadily thereafter.

4.6 Generating functions prove congruences

In this section we give one or two examples of the power of the generating function method in proving congruences among combinatorial numbers. A congruence between two generating functions means that the congruence holds between every pair of their corresponding coefficients.

Example 1. Stirling numbers of the first kind

We found, in chapter 3, that the Stirling numbers of the first kind $\begin{bmatrix} n \\ k \end{bmatrix}$ have the generating function

$$\sum_k \begin{bmatrix} n \\ k \end{bmatrix} x^k = x(x+1)(x+2) \cdots (x+n-1). \tag{4.6.1}$$

Suppose we are interested in finding some criterion for deciding the evenness or oddness of these numbers.

If we read (4.6.1) modulo 2, it becomes

$$\sum_k \begin{bmatrix} n \\ k \end{bmatrix} x^k \equiv x(x+1)x(x+1) \cdots \qquad (\text{mod } 2)$$
$$= x^{\lceil n/2 \rceil}(x+1)^{\lfloor n/2 \rfloor}. \tag{4.6.2}$$

Now take the coefficient of x^k on both sides, and find that

$$\begin{bmatrix} n \\ k \end{bmatrix} \equiv [x^k] x^{\lceil n/2 \rceil}(x+1)^{\lfloor n/2 \rfloor} \qquad (\text{mod } 2)$$
$$= \left[x^{k-\lceil n/2 \rceil} \right] (1+x)^{\lfloor n/2 \rfloor} \tag{4.6.3}$$
$$= \binom{\lfloor n/2 \rfloor}{k - \lceil n/2 \rceil}.$$

Theorem 4.6.1. *The Stirling number $\begin{bmatrix} n \\ k \end{bmatrix}$ has the same parity as the binomial coefficient $\binom{\lfloor n/2 \rfloor}{k-\lceil n/2 \rceil}$. In particular, $\begin{bmatrix} n \\ k \end{bmatrix}$ is an even number if $k < \lceil n/2 \rceil$.*
∎

Example 2. The other Stirling numbers

In the case of the Stirling numbers that count set partitions, the $\left\{ \begin{smallmatrix} n \\ k \end{smallmatrix} \right\}$'s, we found in (1.6.5) that they have the ops generating function

$$\sum_n \left\{ \begin{matrix} n \\ k \end{matrix} \right\} x^n = \frac{x^k}{(1-x)(1-2x) \cdots (1-kx)}.$$

Again, suppose we read the equation modulo 2. Then we would find that

$$\sum_n \left\{ \begin{matrix} n \\ k \end{matrix} \right\} x^n \equiv \frac{x^k}{(1-x)^{\lceil k/2 \rceil}} \qquad (\text{mod } 2)$$
$$= x^k \sum_h \binom{\lceil k/2 \rceil + h - 1}{h} x^h.$$

Now take the coefficient of x^n throughout. The result is:

Theorem 4.6.2. *The Stirling number $\left\{n \atop k\right\}$ has the same parity as the binomial coefficient*

$$\binom{\lceil k/2 \rceil + n - k - 1}{n - k}. \tag{4.6.4}$$

4.7 The cycle index of the symmetric group

We have already studied the Stirling numbers of the first kind, which give the number of permutations of n letters that have exactly k cycles. Now we'll look for much more detailed information about the cycles of permutations. Instead of considering only the *number* of cycles that a permutation has, we will be interested in the numbers of cycles that it has *of each length*.

So let $a = \{a_1, a_2, a_3, \ldots\}$ be a given sequence of nonnegative integers for which $n = a_1 + 2a_2 + 3a_3 + \cdots$ is finite. How many permutations of n letters have exactly a_1 cycles of length 1 and exactly a_2 cycles of length 2 and etc.? For a given permutation σ, we will call the vector $a = a(\sigma)$ the *cycle type* of σ. It tells us the numbers of cycles of each length that σ has.

Let $c(a)$ denote the required number of permutations, and write

$$\phi_n(x) = \sum_{\substack{a_1 + 2a_2 + \cdots = n \\ a_1 \geq 0, a_2 \geq 0, \ldots}} c(a) x_1^{a_1} x_2^{a_2} \cdots. \tag{4.7.1}$$

Then $\phi_n(x)$ is called *the cycle index of the symmetric group S_n.* If we can somehow find $\phi_n(x)$ then the coefficient of each monomial x^a is the number of permutations of n letters whose cycle type is a.

We are going to find the "grand" generating function

$$C(x, t) = \sum_{n=1}^{\infty} \phi_n(x) \frac{t^n}{n!} \tag{4.7.2}$$

which will turn out to have a surprisingly elegant form (see (4.7.5) below), considering the large amount of information that it contains.

The derivation will be unusual in at least one respect. Most often a generating function is a way-station on the road to finding an exact formula for something. But in this problem we will *begin* by finding an exact formula for $c(a)$. It will then be easy to check that its generating function really generates the sequence.

Well, for a given a, how many permutations σ have a for their cycle type? We will first prove a lemma, and then give the answer.

Lemma A. *Given integers* m, a, k. *The number of ways of choosing* ka
letters from m *(distinct) given letters, and arranging them into* a *cycles of*
length k *is*

$$f(m, a, k) = \frac{m!}{(m - ka)! k^a a!}. \qquad (4.7.3)$$

Proof. First choose an ordered ka-tuple of the letters, which can be done
in $m!/(m - ka)!$ ways. Then arrange each consecutive block of k letters in a
cycle, which gives us our set of a cycles. However, we claim that every fixed
set of a cycles of length k will arise exactly $k^a a!$ times in this construction.
Indeed, that set will occur in every ordering of the list of cycles ($a!$ such).
Furthermore, the same set of a cycles results from each of the k possible
circular permutations of elements within blocks of k consecutive entries in
the original ka-tuple, i.e., k^a times. ∎

Hence if we are given n letters, and a sequence of nonnegative integers
$a_1, a_2, \ldots$ such that

$$a_1 + 2a_2 + 3a_3 + \cdots = n,$$

then the number of ways of forming these letters into a_1 1-cycles and a_2
2-cycles, and $\ldots$, is evidently

$$f(n, a_1, 1) f(n - a_1, a_2, 2) f(n - a_1 - 2a_2, a_3, 3) \cdots$$
$$= \left(\frac{n!}{(n - a_1)! 1^{a_1} a_1!} \right) \left(\frac{(n - a_1)!}{(n - a_1 - 2a_2)! 2^{a_2} a_2!} \right) \cdots \qquad (4.7.4)$$
$$= \frac{n!}{a_1! a_2! \cdots 1^{a_1} 2^{a_2} 3^{a_3} \cdots},$$

where Lemma A was used. This yields the following explicit formula for
the number of permutations of each given cycle type.

Theorem 4.7.1. *Let* $\boldsymbol{a}$ *be nonnegative integers for which* $\sum_j j a_j = n$.
Then the number of permutations of n *letters that have* $\boldsymbol{a}$ *for their cycle*
type is exactly

$$c(\boldsymbol{a}) = \frac{n!}{\prod_{j \geq 1} (a_j! j^{a_j})}.$$

Next we'll look for the generating function of the quantities $c(\boldsymbol{a})$. Since
there are infinitely many variables $\boldsymbol{a}$ we shouldn't be surprised by the need

for a generating function in infinitely many variables. We calculate

$$
\begin{aligned}
C(\boldsymbol{x}, t) &= \sum_{n \geq 0} \frac{\phi_n(\boldsymbol{x})}{n!} t^n \\
&= \sum_{n \geq 0} \frac{t^n}{n!} \sum_{\substack{a_1 + 2a_2 + \cdots = n \\ a_1 \geq 0, a_2 \geq 0, \ldots}} c(\boldsymbol{a}) \boldsymbol{x}^{\boldsymbol{a}} \\
&= \sum_{n \geq 0} \frac{t^n}{n!} \sum_{\substack{a_1 + 2a_2 + \cdots = n \\ a_1 \geq 0, a_2 \geq 0, \ldots}} c(\boldsymbol{a}) x_1^{a_1} x_2^{a_2} \cdots \\
&= \left(\sum_{a_1 \geq 0} \frac{(tx_1)^{a_1}}{1^{a_1} a_1!} \right) \left(\sum_{a_2 \geq 0} \frac{(t^2 x_2)^{a_2}}{2^{a_2} a_2!} \right) \cdots \\
&= e^{tx_1} e^{t^2 x_2/2} e^{t^3 x_3/3} \cdots \\
&= \exp \Big(\sum_{j \geq 1} \frac{x_j t^j}{j} \Big).
\end{aligned}
$$

We have proved the following result.

Theorem 4.7.2. *The coefficient of $t^n/n!$ in*

$$
C(\boldsymbol{x}, t) = \exp \Big(\sum_{j \geq 1} \frac{x_j t^j}{j} \Big) \tag{4.7.5}
$$

is the cycle index of S_n, i.e., the generating function $\phi_n(\boldsymbol{x})$ in (4.7.1) above, of the numbers of permutations of n letters that have each possible cycle type. In more detail, the coefficient of $\boldsymbol{x}^{\boldsymbol{a}} t^n/n!$ is the number of permutations of n letters whose cycle type is $\boldsymbol{a}$. ∎

If this result rather reminds you of the exponential formula, and if you suspect that there must be some connection, you are quite correct. The result of exercise 22 of the previous chapter is a generalization of theorem 4.7.2 to exponential families. Indeed, the theorem is an immediate special case of the result of that exercise, but we thought it might be interesting to give an elementary proof also.

Thus the generating function $C(\boldsymbol{x}, t)$ in (4.7.5) generates the cycle indexes of all of the symmetric groups. We will now give some of its applications to the probabilistic theory of permutations.

The polynomials $\phi_n(\boldsymbol{x})$ of (4.7.1) have coefficients that give the number of permutations of n letters with given cycle type. If we divide them by $n!$, as in (4.7.2), then since $n!$ is the total number of permutations of n letters, we will then be finding the *probabilities* that a permutation has various cycle type vectors. Thus

$$C(\boldsymbol{x}, t) = \sum_n \frac{\phi_n(\boldsymbol{x})}{n!} t^n$$

$$= \sum_n p_n(\boldsymbol{x}) t^n$$

(4.7.6)

where

$$p_n(\boldsymbol{x}) = \sum_{\substack{a_1+2a_2+\cdots=n \\ a_1 \geq 0, a_2 \geq 0, \ldots}} \text{Prob}(\boldsymbol{a}, n) \boldsymbol{x}^{\boldsymbol{a}}$$

(4.7.7)

and $\text{Prob}(\boldsymbol{a}, n)$ is the probability that a permutation of n letters has the cycle type $\boldsymbol{a}$.

Just ahead of us now there lie some very pretty theorems. They are theorems that give very quantitative answers to questions that don't seem to have any quantities in them. For instance, take this question, which is a simple illustration of the genre: *what is the probability that a permutation has no fixed points?*

Notice that the question doesn't tell us how many letters the permutation permutes. There is no 'n' in the question. But it has a nontrivial answer: $1/e$, as we discovered in (4.2.10). To interpret a question like this, one proceeds as follows. Let $f(n)$ be the number of permutations of n letters that have no fixed points. Then $f(n)/n!$ is the probability that a permutation of n letters has no fixed points. Since, in this case, $\lim_{n \to \infty} f(n)/n!$ exists and is equal to $1/e$, we can then say that, in this precise sense, *the probability that a permutation has no fixed points is $1/e$.*

There are many lovely questions about permutations that sound like *what is the probability that a permutation has ...?*, in which the number of letters that the permutations act upon is not even mentioned, and to which the answers are nontrivial numbers, like the $1/e$ above. We are about to derive handfuls of them at once. But first we need a lemma.

Lemma B. *Let $\sum_j b_j$ be a convergent series. Then in the power series expansion of the function*

$$\frac{1}{1-t} \sum_j b_j t^j = \sum_n \alpha_n t^n,$$

we have that $\lim_n \alpha_n$ exists and is equal to $\sum_j b_j$.

Proof. By Rule 5 of section 2.2, α_n is the sum of those b_j for which $j \leq n$. The latter sum clearly approaches the limit stated. ∎

Now let S be a (finite or infinite) set of positive integers, with the property that

$$\sum_{s \in S} \frac{1}{s} < \infty.$$

In the generating function $C(\boldsymbol{x}, t)$ we set all $x_i = 1$ for $i \notin S$. That means that we are declaring ourselves to have no interest in any cycle lengths other than those in S. The others can be whatever they please. Then C becomes

$$C(\boldsymbol{x}, t) = \exp\left(\sum_{i \in S} x_i \frac{t^i}{i} + \sum_{i \notin S} \frac{t^i}{i}\right)$$

$$= \exp\left(\sum_{i \in S} \frac{(x_i - 1)t^i}{i} + \log\frac{1}{1-t}\right)$$

$$= \frac{1}{1-t} \exp\left(\sum_{i \in S} \frac{(x_i - 1)t^i}{i}\right).$$

By Lemma B above, the coefficient of t^n in this last expression approaches the limit

$$\exp\left(\sum_{i \in S} \frac{(x_i - 1)}{i}\right)$$

as $n \to \infty$, which proves the following.

Theorem 4.7.3. *Let S be a set of positive integers for which $\sum_{s \in S} 1/s$ converges, and let $\boldsymbol{a}$ be a fixed cycle type vector. The probability that the cycle type vector of a random permutation agrees with $\boldsymbol{a}$ in all of its components whose subscripts lie in S exists and is equal to*

$$e^{-\left(\sum_{s \in S} 1/s\right)} [\boldsymbol{x}^{\boldsymbol{a}}] \exp\left(\sum_{s \in S} \frac{x_s}{s}\right) = \frac{1}{\prod_{s \in S}\left(e^{\frac{1}{s}} s^{a_s} a_s!\right)}. \qquad (4.7.8)$$

As a first example, take $S = \{1\}$, so we are interested only in fixed points. Then from (4.7.8), the probability that a random permutation has exactly a fixed points is $1/(a!e)$, for $a \geq 0$.

Take $S = \{r\}$. Then we see at once that the probability that a random permutation has exactly a r-cycles is $1/(e^{\frac{1}{r}} r^a a!)$, for each $a = 0, 1, \ldots$.

Let $S = \{r, s\}$. The probability that a random permutation has exactly a_r r-cycles and exactly a_s s-cycles is therefore

$$\frac{e^{-1/r - 1/s}}{r^{a_r} s^{a_s} r! s!}.$$

OK, now blindfold yourself, reach into a bag that contains every permutation in the world, and pull one out. What is the probability that none of its cycle lengths is the square of an integer? We claim that the probability is $e^{-\pi^2/6} = .193025\ldots$, so the odds are about 5 to 1 against this event. Indeed, this is just the case where we take S to be the set of all squares of integers, and use the fact that $1 + 1/2^2 + 1/3^2 + \cdots = \pi^2/6$.

For a final example, what is the probability that a randomly chosen permutation contains equal numbers of 1-cycles and 2-cycles? Well, if that number is j, then the probability is $e^{-3/2}/(2^j j!^2)$, for each $j = 0, 1, 2, \ldots$. If we sum over all of these j, we find that the required probability is

$$e^{-3/2} \sum_{j=0}^{\infty} \frac{1}{2^j j!^2} = 0.34944033\ldots.$$

We can recast the result in theorem 4.7.3 in the language of the Poisson distribution. The Poisson distribution is a probability distribution on the nonnegative integers $j = 0, 1, 2, \ldots$ that occurs very naturally in a number of areas of application, such as in the theory of waiting lines. It is given by

$$\mathrm{Prob}(j) = e^{-M} \frac{M^j}{j!} \qquad (j = 0, 1, 2, \ldots)$$

where M is the mean.

Theorem 4.7.3 then asserts the following: If a set S is fixed, for which $\sum_{s \in S} 1/s < \infty$, then for a randomly chosen permutation the numbers of cycles of each length $s \in S$ have asymptotically independent Poisson distributions in which the mean number of s-cycles is $1/s$ for each $s \in S$.

4.8 How many permutations have square roots?

Let σ be a permutation. There may or may not be a permutation τ such that $\sigma = \tau^2$. We want to describe and to count the σ's that do have square roots, in this sense. More generally, σ has a kth root if there is a τ such that $\sigma = \tau^k$, and again, we would like to know the number of permutations of n letters that have kth roots.

The answers to these questions involve some fairly spectacular generating functions, and the methods will lean strongly on the cycle index results of the previous section, and in particular on theorem 4.7.2.

We begin with the square root problem. So let τ be a permutation, and consider a single cycle of τ, say this one:

$$8 \to 3 \to 13 \to 19 \to 7 \to 12 \to 8.$$

What happens to that cycle when we square τ? The mapping τ^2 executes the permutation τ twice, so it carries 8 into 13 and 13 into 7, etc. Thus we go hopping around the cycle of τ, visiting every second member, until we return to our starting point. A cycle of even length therefore falls apart into two cycles of half the length, while an odd cycle remains a cycle of the same length, although it becomes a different cycle of that length.

In the example above, the cycle shown breaks into two cycles, like this:

$$8 \to 13 \to 7 \to 8 \qquad \text{and} \qquad 3 \to 19 \to 12 \to 3.$$

In general, every cycle of τ whose length is $2m$ will contribute two cycles of length m to τ^2.

A cycle of even length in τ^2, therefore, can only be the result of splitting a cycle of twice its length in τ into two cycles. Hence, if σ has a square root, then *the number of cycles that it has of each even length must be even.*

Conversely, let σ be any permutation that has this property. Then we claim that $\sigma = \tau^2$ for at least one τ. In fact we can construct such a τ (in how many ways?). To do that, pick up a pair of cycles of the same even length and thread them together into a single cycle of twice that length, as in the two examples above, by taking alternately a letter from one of the cycles and a letter from the other. These threaded cycles are all parts of the permutation τ that is being constructed.

What do we do with the odd cycles of σ? If we are given a cycle of odd length $2m + 1$ in σ we convert it into a cycle of the same odd length in τ as follows. Let the letters in the given cycle of σ be

$$a_1 \to a_2 \to a_3 \to \cdots \to a_{2m} \to a_{2m+1} \to a_1.$$

Then into τ we put the cycle

$$a_1 \to a_{m+2} \to a_2 \to a_{m+3} \to a_3 \to a_{m+4} \to \cdots \to a_{2m+1} \to a_{m+1}.$$

This cycle clearly has the property that if we square it then it will be back in the original order as in σ, and completes the proof of the following theorem (as well as giving us an algorithm for finding the square root of a permutation!).

Theorem 4.8.1. *A permutation σ has a square root if and only if the numbers of cycles of σ that have each even length are even numbers.* ∎

Now let $f(n, 2)$ be the number of permutations of n letters that have square roots. We seek the generating function of the sequence. Consider the cycle type vector $\boldsymbol{a}$ of such a permutation. The even-indexed components must be even numbers, and the odd-indexed components are arbitrary.

According to theorem 4.7.2, the coefficient of $\boldsymbol{x}^{\boldsymbol{a}} t^n / n!$ in the product

$$e^{x_1 t} e^{x_2 t^2 / 2} e^{x_3 t^3 / 3} \ldots$$

is the number of permutations of n letters whose cycle type is $\boldsymbol{a}$. The sum of these coefficients over all of the cycle types that we are considering, namely $\boldsymbol{a}$'s for which the even-indexed entries are even, is obtained as follows. In the product of the exponential functions above, put $x_1 = 1$, because all values of a_1 are admissible. In the second exponential factor, $e^{x_2 t^2 / 2}$, don't use the whole exponential series. Since only even powers of x_2 are admissible, use the subseries of even powers of the exponential series, namely the cosh series, and then put $x_2 = 1$. Then put $x_3 = 1$. Then use the $\cosh (x_4 t^4 / 4)$ series and put $x_4 = 1$, and so forth.

The result will be that the number $f(n, 2)$ of permutations of n letters that have square roots satisfies

$$\sum_{n \geq 0} f(n, 2) \frac{t^n}{n!} = e^t \cosh{(t^2/2)} e^{t^3/3} \cosh{(t^4/4)} e^{t^5/5} \cdots$$

$$= \exp{(t + t^3/3 + t^5/5 + \cdots)} \prod_{m \geq 1} \cosh{\left(\frac{t^{2m}}{2m}\right)}$$

$$= \sqrt{\frac{1+t}{1-t}} \prod_{m \geq 1} \cosh{\left(\frac{t^{2m}}{2m}\right)} \qquad (4.8.1)$$

$$= 1 + t + \frac{t^2}{2!} + 3\frac{t^3}{3!} + 12\frac{t^4}{24} + 60\frac{t^5}{5!} + \cdots.$$

Hence the sequence $\{f(n, 2)\}$ begins as

$$1, 1, 1, 3, 12, 60, 270, 1890, 14280, \ldots.$$

Corollary. *Let $p(n)$ be the probability that a permutation of n letters has a square root. Then for each $n = 0, 1, 2, \ldots$, we have $p(2n) = p(2n + 1)$.*

Proof. The generating function in (4.8.1) is of the form $1/(1-t)$ times an even function of t. Hence $f(n, 2)/n!$ is the nth partial sum of the coefficient sequence of an even function. ∎

No bijective proof of this corollary is known. It seems that, more generally, the sequence of probabilities is decreasing, i.e.,

$$p(0) = p(1) \geq p(2) = p(3) \geq p(4) = p(5) \geq \cdots,$$

but this is unproved.

Now let's try the question of kth roots. We let $f(n, k)$ be the number of permutations of n letters that have kth roots, and we seek the egf of $\{f(n, k)\}_{n \geq 0}$.

First, if k is given, for which permutations σ is it true that there exists a permutation τ such that $\sigma = \tau^k$? The generalization of theorem 4.8.1 is the following.

Theorem 4.8.2. *A permutation σ has a kth root if and only if for every $m = 1, 2, \ldots$ it is true that the number of m-cycles that σ has is a multiple of $\gcd(m, k)$.*

To prove this we observe first that if we raise a permutation τ to the kth power, then an r-cycle of τ falls apart into $\gcd(r, k)$ cycles, each of length $r/\gcd(r, k)$. Hence if we consider a particular m-cycle of $\sigma = \tau^k$, then that cycle could have come from an r-cycle of τ, where $m = r/\gcd(r, k)$. Thus $r = m \cdot \gcd(r, k)$, which implies that $r = md$ where d is some divisor of

k. But a cycle of length md of τ would contribute $\gcd(md, k)$ m-cycles to τ^k, which is certainly a multiple of $\gcd(m, k)$, so the claimed condition is necessary.

To show that it is also sufficient, let σ be a permutation that satisfies the condition. We will construct a kth root, τ, of σ. Fix m, and write $g = \gcd(m, k)$. Then the number of m-cycles of σ is a multiple of g, so we can tie them up into bundles of g m-cycles each, and then for each bundle, we can construct a single new cycle of length mg, in such a way that when raised to the kth power that mg-cycle would fall apart into exactly the g m-cycles that were in the bundle. The required kth root τ would then be the union of the mg-cycles thus constructed.

It remains to say how, for a given bundle of g m-cycles, we construct such a single mg-cycle. That is done by weaving together the cycles in the bundle, as follows. Construct a circle with mg places marked consecutively around it. Take the first m-cycle in the bundle and arrange its elements in the marked places, consecutive elements being spaced apart by g places. Then do the same for the second m-cycle in the bundle, etc., and the proof is complete. ∎

To obtain the egf of the sequence $\{f(n, k)\}$ we proceed as in (4.8.1) above. It will be convenient to have a name for the subseries of the exponential series that occur. So let us write $\exp_q(x)$ for the subseries of the exponential series e^x that is obtained by choosing only the powers of x that are divisible by q. That is

$$\exp_q(x) = \sum_{j \geq 0} \frac{x^{jq}}{(jq)!} \qquad (q = 1, 2, 3, \ldots).$$

Thus $\exp_1(x) = e^x$, $\exp_2(x) = \cosh x$, $\exp_3(x)$ is explicitly shown in eq. (2.4.7) of chapter 2, etc.

Now following the argument that led to (4.8.1) we obtain this generalization.

Theorem 4.8.3. *Let $f(n, k)$ be the number of permutations of n letters that have a kth root. Then we have*

$$\sum_{n=0}^{\infty} f(n, k) \frac{x^n}{n!} = \prod_{m=1}^{\infty} \exp_{\gcd(m,k)} \left(\frac{x^m}{m} \right) \qquad (k = 1, 2, 3, \ldots). \qquad (4.8.2)$$

The reader is invited to check that this reduces to a triviality when $k = 1$, and to (4.8.1) when $k = 2$.

A short table of $f(n, k)$ ($1 \leq n \leq 10; 2 \leq k \leq 7$) is shown below.

$k = 2$:	1	1	3	12	60	270	1890	14280	128520	1096200
$k = 3$:	1	2	4	16	80	400	2800	22400	181440	1814400
$k = 4$:	1	1	3	12	60	270	1890	13020	117180	1039500
$k = 5$:	1	2	6	24	96	576	4032	32256	290304	2612736
$k = 6$:	1	1	1	4	40	190	1330	8680	52920	340200
$k = 7$:	1	2	6	24	120	720	4320	34560	311040	3110400

4.9 Counting polyominoes

By a *cell* we will mean the interior and boundary of a unit square in the x-y plane, if the vertices of the square are at lattice points (points whose coordinates are both integers). Let P be a collection of cells. We associate with P a graph, whose vertices correspond to the cells of P, and in which two vertices are joined by an edge in the graph if the two cells to which they correspond intersect in a line segment (rather than in a vertex, or not at all). We say that P is a connected collection of cells if the graph associated with P is a connected graph.

A collection P of cells is in *standard position* if all of its cells lie in the first quadrant, and at least one of them intersects the y axis and at least one of them intersects the x axis.

A *polyomino* is a connected collection of cells that is in standard position.

Here are all of the polyominoes that have one, two, or three cells:

Sometimes polyominoes are called *animals*. This is because one can imagine a single cell that 'grows' by sprouting a new cell along one of its edges. Then that two-celled animal would grow a new cell along one of its edges, etc. If $f(n)$ is the number of n-celled polyominoes, then from the picture above we see that $\{f(n)\} = 1, 2, 6, \ldots$. It would be good to be able to say that in this section we are going to derive the generating function etc. for the sequence $f(n)$. We aren't going to do that, though. The sequence and its generating function are unknown, despite a great deal of effort that has been invested in the problem.

Various special kinds of polyominoes, however, have been counted, with respect to various properties of the polyomino. For instance, among the properties that a polyomino has, one might mention its *area*, or number of cells, and its *perimeter*. So one might ask for the number of polyominoes of some special kind whose area is n, or the number whose perimeter is m, or the number whose area is n and whose perimeter is m, or the generating functions of any of these, etc. For a survey of recent progress in such questions see [De1] and [De2].

What we will do in this section will be to count a special kind of polyomino that is called *horizontally convex* (HC). An HC-polyomino is one in which every row is a single contiguous block of cells. The picture below shows a typical HC-polyomino.

Another special kind of polyomino is called *convex*. A polyomino is convex if it is both vertically and horizontally convex. One of the striking results in the theory of polyominoes is the fact that there are exactly

$$(2n + 11)4^n \quad 4(2n + 1)\binom{2n}{n} \qquad (4.9.1)$$

convex polyominoes of perimeter $2n + 8$. The exact number of convex polyominoes of area n is unknown.

There are interesting problems involved in counting HC-polyominoes either by area or by perimeter. We are going to count them here by area. It is worth noting that the question of enumerating them by perimeter has also been solved [De2], and the solution involves a remarkable generating function, which looks like this: if c_n is the number of HC-polyominoes whose perimeter is $2n + 2$ then

$$\sum_{n \geq 0} c_n t^n = \frac{\sqrt{-(AC^{1/3} + D + EC^{-1/3}) - F}}{2\sqrt{AH}} - \frac{H}{2\sqrt{2A}} - G$$

in which $A, B, \ldots, H$ are certain specific functions. For instance $A = 18t^4(2t^3 - 23t^2 + 38t - 18)^2$. For the complete list see [De2].

We return to the problem of counting HC-polyominoes by area, which is similar to the enumeration of 'fountains' in section 2.2, and our method of attack will be similar.

Let $f(n, k, t)$ be the number of HC-polyominoes of n cells, having k rows, of which t are in the top row. If we strip off the top row of one of these polyominoes, what will remain will have $n - t$ cells, arranged in $k - 1$ rows, with some number $r \geq 1$ in the top row. Hence after removing the top row, there are $f(n - t, k - 1, r)$ possibilities for what remains, for some r. However, each one of those possibilities generates $r + t - 1$ of the original

(n, k, t) HC-polyominoes, by adjoining a top row of t cells, and sliding it left and right through all legal positions atop the second row.

Hence we have

$$f(n, k, t) = \sum_{r \geq 1} f(n - t, k - 1, r)(r + t - 1) \qquad (k \geq 2;\ f(n, 1, t) = \delta_{t,n}).$$

$$(4.9.2)$$

If we define the generating functions $F_{k,t}(x) = \sum_n f(n, k, t)x^n$, then we have $F_{1,t}(x) = x^t$, for $t \geq 1$ and after multiplying (4.9.2) by x^n and summing over n, we obtain

$$F_{k,t}(x) = x^t \sum_{r \geq 1} (r + t - 1)F_{k-1,r}(x) \qquad (k \geq 2). \qquad (4.9.3)$$

Now let $U_k(x) = \sum_{r \geq 1} F_{k,r}(x)$ and $V_k(x) = \sum_{r \geq 1} r F_{k,r}(x)$. Then $U_1(x) = x/(1 - x)$ and $V_1(x) = x/(1 - x)^2$. Further, from (4.9.3),

$$F_{k,t}(x) = x^t(V_{k-1}(x) + (t - 1)U_{k-1}(x)) \qquad (k \geq 2), \qquad (4.9.4)$$

and if we sum on t we find that

$$U_k(x) = \frac{x}{1 - x}V_{k-1}(x) + \frac{x^2}{(1 - x)^2}U_{k-1}(x) \qquad (k \geq 2). \qquad (4.9.5)$$

If we first multiply (4.9.4) by t and then sum on t we find

$$V_k(x) = \frac{x}{(1 - x)^2}V_{k-1}(x) + \frac{2x^2}{(1 - x)^3}U_{k-1}(x) \qquad (k \geq 2). \qquad (4.9.6)$$

We now have two simultaneous recurrences to solve for the sequences U_k and V_k. To do that we eliminate the V_k sequence as follows: solve (4.9.5) for V_{k-1} in terms of U_k and U_{k-1}, and substitute the result in (4.9.6). After simplification we obtain a single three term recurrence for the U's, viz.

$$\frac{1 - x}{x}U_{k+1}(x) - \frac{x + 1}{1 - x}U_k(x) - \frac{x^2}{(1 - x)^3}U_{k-1}(x) = 0 \qquad (k \geq 1), \quad (4.9.7)$$

along with the initial data $U_0(x) = 0$ and $U_1(x) = x/(1 - x)$.

Finally, to solve (4.9.7) we introduce the generating function $\phi(x, y) = \sum_{k \geq 0} U_k(x)y^k$. Then, if we multiply (4.9.7) by y^k and sum over $k \geq 1$ we get

$$\frac{1 - x}{xy}\left\{\phi(x, y) - U_1(x)y\right\} - \frac{x + 1}{1 - x}\phi(x, y) - \frac{x^2 y}{(1 - x)^3}\phi(x, y) = 0.$$

If we use the initial conditions and solve for ϕ, the result is that

$$\phi(x,y) = \sum_{k \geq 0} U_k(x)y^k = \sum_{n,k,r} f(n,k,r)x^n y^k$$

$$= \frac{xy(1-x)^3}{(1-x)^4 - xy(1-x-x^2+x^3+x^2y)}. \qquad (4.9.8)$$

Notice that the sum over r has no variable attached to it; it acts directly on $f(n,k,r)$ and yields the number of HC-polyominoes of n cells and k rows, without regard to how many cells are in the top row. Thus if $g(n,k)$ is that number, then

$$\sum_{n,k} g(n,k)x^n y^k = \frac{xy(1-x)^3}{(1-x)^4 - xy(1-x-x^2+x^3+x^2y)}. \qquad (4.9.9)$$

For the complete 3-variable generating function of the sequence $\{f(n,k,r)\}$, see exercise 21 at the end of this chapter.

Perhaps we are interested only in the total number of HC-polyominoes, and we don't need to know the number of rows. In that case we let $y = 1$ in (4.9.8) and we find the following result, which is due to D. Klarner, who used different methods.

Theorem 4.9.1. *Let $f(n)$ be the number of n-celled HC-polyominoes. Then*

$$\sum_{n \geq 1} f(n)x^n = \frac{x(1-x)^3}{1 - 5x + 7x^2 - 4x^3}$$

$$= x + 2x^2 + 6x^3 + 19x^4 + 61x^5 + 196x^6 + 629x^7 + 2017x^8$$

$$+ 6466x^9 + 20727x^{10} + 66441x^{11} + 212980x^{12} + \cdots$$

$$(4.9.9)$$

We now will give a preview of the material in chapter 5, by working out an *asymptotic formula* for $f(n)$. To do this we take the generating function in (4.9.9) and expand it in partial fractions. This gives

$$\frac{x(1-x)^3}{1 - 5x + 7x^2 - 4x^3} = -\frac{5}{16} + \frac{x}{4} + \frac{c_1}{1 - \xi_1 x} + \frac{c_2}{1 - \xi_2 x} + \frac{c_3}{1 - \xi_3 x},$$

in which $\xi_1 = 3.20556943...$ and $\xi_{2,3} = 0.897215 \pm .665457i$.

Thus the number of HC-polyominoes is, for $n \geq 2$,

$$f(n) = c_1 \xi_1^n + c_2 \xi_2^n + c_3 \xi_3^n$$

$$= c_1 \xi_1^n + O(|\xi_2|^n)$$

$$= 0.1809155018\ldots (3.2055694304\ldots)^n + O(1.1171^n).$$

The first term of this formula gives, for example, $f(12) = 212979.61$, compared to 212980, the exact value, as shown in (4.9.9).

4.10 Exact covering sequences

Every positive integer n is either 1 mod 2 or 0 mod 4 or 2 mod 4, as a moment's reflection will confirm. So the three pairs $(a_1, b_1) = (1, 2)$, $(a_2, b_2) = (0, 4)$ and $(a_3, b_3) = (2, 4)$ of residues and moduli *exactly cover* the positive integers.

An *exact covering sequence* (ECS) is a set (a_i, b_i) $(i = 1, \ldots, k)$ of ordered pairs of nonnegative integers with the property that for every nonnegative integer n there is one and only one i such that $1 \le i \le k$ and $n \equiv a_i \bmod b_i$.

In this section we will give the basic theory of such sequences and deal, in two or three different ways, with the question of how we can tell if a given sequence of pairs is or is not an exact covering sequence.

Here is what generating functions have to contribute to this subject.

Suppose (a_i, b_i) $(i = 1, \ldots, k)$ is an exact covering sequence. Then in the series

$$\sum_{i=1}^{k} \sum_{t \ge 0} x^{a_i + t b_i}$$

every nonnegative integer n occurs exactly once as an exponent of x, so it must be true that the series shown is equal to $1/(1-x)$. If we perform the summation over t, we find that

$$\sum_{i=1}^{k} \frac{x^{a_i}}{1 - x^{b_i}} = \frac{1}{1 - x}. \tag{4.10.1}$$

For example

$$\frac{x}{1 - x^2} + \frac{1}{1 - x^4} + \frac{x^2}{1 - x^4} = \frac{1}{1 - x}.$$

Theorem 4.10.1. *For a set of pairs (a_i, b_i) $(i = 1, \ldots, k)$ to be an exact covering sequence it is necessary and sufficient that the relation (4.10.1) hold.*

One conclusion that we can draw immediately is that in an ECS we must have $\sum_i 1/b_i = 1$. To see that, just multiply (4.10.1) by $1 - x$ and let $x \to 1$. But we can learn much more by comparing the partial fraction expansions of the left and right sides of (4.10.1).

For the left side, we have

$$\sum_{i=1}^{k} \frac{x^{a_i}}{1 - x^{b_i}} = \sum_{\omega : \omega^N = 1} \frac{A(\omega)}{\omega - x}$$

where

$$A(\omega) = \lim_{x \to \omega} (\omega - x) \sum_{i=1}^{k} \frac{x^{a_i}}{1 - x^{b_i}}$$

$$= \sum_{j:\omega^{b_j}=1} \frac{\omega^{a_j+1}}{b_j},$$

and $N = \text{l.c.m.}\{b_j\}$. If we compare with the right side of (4.10.1) we see that the $A(\omega)$'s must all vanish except that $A(1) = 1$. Hence we must have

$$\sum_{j:\omega^{b_j}=1} \frac{\omega^{a_j}}{b_j} = \begin{cases} 1, & \text{if } \omega = 1; \\ 0, & \text{otherwise.} \end{cases}$$

Now $A(\omega)$ surely vanishes unless ω is a root of unity. So let $\omega = e^{2\pi i r/s}$, where $s \geq 1$ and $(r, s) = 1$, be a primitive sth root of unity. Then our conditions take the form

$$\sum_{j:s\backslash b_j} \frac{\omega^{a_j}}{b_j} = \begin{cases} 1 & \text{if } s = 1; \\ 0 & \text{otherwise.} \end{cases} \tag{4.10.2}$$

Hence, associated with any sequence of pairs $(a_i, b_i)|_{i=1}^{k}$ we can define polynomials

$$\psi_s(z) = \sum_{j:s\backslash b_j} \frac{z^{a_j}}{b_j} \qquad (s = 1, 2, 3, \ldots). \tag{4.10.3}$$

In terms of these polynomials we can restate our conditions (4.10.2) as follows: necessary and sufficient for the given set of pairs to constitute an exact covering sequence is that for each $s > 1$ the polynomial ψ_s should vanish at the primitive sth rots of unity, and $\psi_1(1) = 1$.

However, any polynomial that vanishes at all of the primitive sth roots of unity must be divisible by the cyclotomic polynomial (see section 2.6, example 2)

$$\Phi_s(z) = \prod_{r:(r,s)=1;0<r<s} (z - e^{2\pi i r/s}) \qquad (s = 1, 2, 3, \ldots)$$

since $\Phi_s(z)$ has *only* those roots. The first few cyclotomic polynomials are

$$1 - z, 1 + z, 1 + z + z^2, 1 + z^2, 1 + z + z^2 + z^3 + z^4, 1 - z + z^2, \ldots.$$

If we put all of this together, we obtain the following result.

Theorem 4.10.2. *A set of pairs of integers* $(a_1, b_1), \ldots, (a_k, b_k)$, *in which the a's are nonnegative and the b's are positive, is an exact covering sequence if and only if* $\sum_j 1/b_j = 1$ *and for each* $s > 1$, *the polynomial* $\psi_s(z)$, *of (4.10.3), is divisible by the cyclotomic polynomial* $\Phi_s(z)$. ∎

For an example, take the pairs

$$(0, 4), (2, 4), (1, 6), (3, 6), (5, 12), (11, 12).$$

Then $\sum_j 1/b_j = 1/4 + 1/4 + 1/6 + 1/6 + 1/12 + 1/12 = 1$, and the divisibility conditions of the theorem look like this:

$$
\begin{array}{lll}
\Phi_2(z) = (1 + z) & \text{divides} & \frac{1}{4} + z^2/4 + z/6 + z^3/6 + z^5/12 + z^{11}/12 \\
\Phi_3(z) = (1 + z + z^2) & \text{divides} & z/6 + z^3/6 + z^5/12 + z^{11}/12 \\
\Phi_4(z) = (1 + z^2) & \text{divides} & \frac{1}{4} + z^2/4 + z^5/12 + z^{11}/12 \\
\Phi_6(z) = (1 - z + z^2) & \text{divides} & z/6 + z^3/6 + z^5/12 + z^{11}/12 \\
\Phi_{12}(z) = (1 - z^2 + z^4) & \text{divides} & z^5/12 + z^{11}/12.
\end{array}
$$

These are all readily checked, and so the given pairs are an ECS.

Theorem 4.10.2 has a number of corollaries, some of which are left as exercises. One of them, however, is quite clear. If $B = \max\{b_j\}$, then B cannot occur just once among the moduli $\{b_j\}$. Indeed, if we take $s = B$ in the theorem, we discover that $\psi_B(z)$ must have enough monomials in it to allow it to be divisible by $\Phi_B(z)$, so it must surely have at least two monomials.

Exercises

1. Given a coin whose probability of turning up 'heads' is p, let p_n be the probability that the first occurrence of 'heads' is at the nth toss of the coin. Evaluate p_n and the opsgf of the sequence $\{p_n\}$. Use that opsgf to find the mean of the number of trials till the first 'heads' and the standard deviation of that number.

2. In the *coupon collector's problem* we imagine that we would like to get a complete collection of photos of movie stars, where each time we buy a box of cereal we acquire one such photo, which may of course duplicate one that is already in our collection. Suppose there are d different photos in a complete collection. Let p_n be the probability that exactly n trials are needed in order, for the first time, to have a complete collection.

    (a) Show that

$$p_n = \frac{d!\left\{{n-1 \atop d-1}\right\}}{d^n},$$

       where $\left\{{n \atop k}\right\}$ is the Stirling number of the second kind (see section 1.6).

    (b) Let $p(x) \overset{ops}{\longleftrightarrow} \{p_n\}$. Show that

$$p(x) = \frac{(d-1)!x^d}{(d-x)(d-2x)\cdots(d-(d-1)x)}.$$

    (c) Find, directly from the generating function $p(x)$, the average number of trials that are needed to get a complete collection of all d coupons.

    (d) Similarly, using $p(x)$, find the standard deviation of that number of trials.

    (e) In the case $d = 10$, how many boxes of cereal would you expect to have to buy in order to collect all 10 different kinds of pictures?

3. (*First return times on trees*) By a *random walk* on a graph we mean a walk among the vertices of the graph, which, having arrived at some vertex v, goes next to a vertex w that is chosen uniformly from among the neighbors of v in the graph.

If T is a tree, and v is a vertex of T, let $p(j; v; T)$ denote the probability that a random walk on T which starts at vertex v, returns to v for the first time after exactly j steps.

Now let T_1, T_2 be trees, let v_i be a vertex of T_i for $i = 1, 2$, and let T be the tree that is formed from these two by adding edge (v_1, v_2). Finally, let $F_1(x; v_1)$, $F_2(x; v_2)$, $F(x; v_1)$ be the opsgf's of the sequences $\{p(j; v_1; T_1)\}_{j \geq 0}$, $\{p(j; v_2; T_2)\}_{j \geq 0}$, and $\{p(j; v_1; T)\}_{j \geq 0}$, respectively.

(a) Show that

$$F(x; v_1) = \frac{1}{d_1 + 1} \left\{ d_1 F_1(x; v_1) + \frac{x^2}{d_2 + 1 - d_2 F_2(x; v_2)} \right\},$$

where d_i is the degree of vertex v_i in the tree T_i, for $i = 1, 2$.

(b) Let $\mu_T(a)$ be the *average* number of steps in a random walk that starts at vertex $a \in T$ and stops when it returns to a for the first time. Show, by differentiating the answer to part (a), that

$$\mu_T(v_1) = 2 + \frac{d_1}{d_1 + 1} \mu_{T_1}(v_1) + d_2 \mu_{T_2}(v_2).$$

(c) Let the tree T be a path of $n + 1$ vertices. Show that the mean return time of a walk that begins at vertex v is $2n$ if v is one of the two endpoints, and is n for all other v (surprisingly?).

(d) Again, if T is a path of n vertices, and if $P_n(x)$ denotes the generating function $F(x; v_1)$ of part (a), where v_1 is an endpoint of T, then find an explicit formula for $P_n(t)$.

4. Find, in terms of $N(x)$, the opsgf of the sequences $\{e_{\leq m}\}$ (resp. $\{e_{\geq m}\}$) which count the objects that have *at most* m properties (resp. at least m properties).

5. What chessboard would you use to derive the number of permutations that have no fixed points? Rederive the formula for this number using the chessboard method.

6. (*Bonferroni's inequalities*) In the sieve method, eq. (4.2.6) computes the number of objects that have no properties at all. Suppose the alternating series on the right were cut off after a certain value $t = m$, say. Show that the result would overestimate e_0 if m were even, and underestimate it for m odd.

To do this, show that the sequence

$$\alpha_m = \sum_{r \geq m} (-1)^{r-m} N_r \qquad (m = 0, 1, 2, \ldots)$$

has the opsgf $e_0 + x \sum_{r \geq 0} e_{r+1} (x + 1)^r$, whose coefficients are obviously nonnegative.

7. (*Bonferroni's inequalities, cont.*) Not only is e_0 alternately under- and overestimated by the successive partial sums of its sieve formula, the same is true of every e_k, the number of objects that have exactly k properties. To show this generatingfunctionologically, define, for each $k, t \geq 0$,

$$\gamma(k, t) = (-1)^{t+1} \left\{ e_k - \sum_{j \leq t} (-1)^j \binom{k + j}{j} N_{k+j} \right\}.$$

Then the problem is to show that all $\gamma(k,t) \geq 0$.

To do this,

    (a) let $\Gamma(x,y)$ be the 2-variable opsgf of the γ's. Then multiply the definition of the γ's by $x^k y^t$, sum over $k, t \geq 0$, and show that

$$\Gamma(x,y) = \frac{E(x + (1+y)) - E(x)}{(1+y)}.$$

    (b) It now follows that the γ's are nonnegative, and in fact that

$$\gamma(k,t) = \sum_r \binom{r}{k}\binom{r-k-1}{t} e_r \qquad (k, t \geq 0).$$

8. Show that

$$\sum_r \binom{n}{\lfloor \frac{r}{2} \rfloor} x^r = (1+x)(1+x^2)^n.$$

Then use Snake Oil to evaluate

$$\sum_k \binom{n}{k}\binom{n-k}{\lfloor \frac{m-k}{2} \rfloor} y^k$$

explicitly, when $y = \pm 2$ (due to D. E. Knuth). Find the generating function of these sums, whatever the value of y.

9. Let G be a graph of n vertices, and let positive integers x, λ be given. Let $P(\lambda; x; G)$ denote the number of ways of assigning one of λ given colors to each of the vertices of G in such a way that exactly x edges of G have both endpoints of the same color.

Formulate the question of determining P as a sieve problem with a suitable set of objects and properties. Find a formula for $P(\lambda; x; G)$, and observe that it is a polynomial in the two variables λ and x. The *chromatic polynomial* of G is $P(\lambda; 0; G)$.

10.

    (a) Let w be a word of m letters over an alphabet of k letters. Suppose that no final substring of w is also an initial string of w. Use the sieve method to count the words of n letters, over that alphabet of k letters, that do not contain the substring w.

    (b) Use the Snake Oil method on the sum that you got for the answer in part (a).

11. Use the Snake Oil method to do all of the following:

    (a) Find an explicit formula, not involving sums, for the polynomial

$$\sum_{k \geq 0} \binom{k}{n-k} t^k.$$

(b) Invent a really nasty looking sum involving binomial coefficients that isn't any of the ones that we did in this chapter, and evaluate it in simple form.

(c) Evaluate

$$\sum_k \binom{2n+1}{2p+2k+1}\binom{p+k}{k},$$

and thereby obtain a 'Moriarty identity.'

(d) Show that

$$\sum_m \binom{r}{m}\binom{s}{t-m} = \binom{r+s}{t}.$$

Then evaluate

$$\sum_k \binom{n}{k}^2.$$

(e) Show that (Graham and Riordan)

$$\sum_k \binom{2n+1}{2k}\binom{m+k}{2n} = \binom{2m+1}{2n} \qquad (n \geq 0).$$

(f) Show that for all $n \geq 0$

$$\sum_k \binom{n}{k}\binom{k}{j}x^k = \binom{n}{j}x^j(1+x)^{n-j}.$$

(g) Show that for all $n \geq 0$

$$x\sum_k \binom{n+k}{2k}\left(\frac{x^2-1}{4}\right)^{n-k} = \left(\frac{x-1}{2}\right)^{2n+1} + \left(\frac{x+1}{2}\right)^{2n+1}.$$

(h) Show that for $n \geq 1$

$$\sum_{k\geq 1}\binom{n+k-1}{2k-1}\frac{(x-1)^{2k}x^{n-k}}{k} = \frac{(x^n-1)^2}{n}.$$

12. The Snake Oil Method works not only on sums that involve binomial coefficients, but on all sorts of counting numbers, as this exercise shows.

(a) Let $\{a_n\}$ and $\{b_n\}$ be two sequences whose egf's are, respectively, $A(x)$, $B(x)$. Suppose that the sequences are connected by

$$b_n = \sum_k \begin{bmatrix} n \\ k \end{bmatrix} a_k \qquad (n \geq 0),$$

where the $[\ ]$'s are the Stirling numbers of the first kind. Show that their egf's are connected by

$$B(x) = A\left(\log \frac{1}{(1-x)}\right).$$

(b) Let $\tilde{b}_n$ be the number of ordered partitions of $[n]$ (see (5.2.7)). Show that

$$\sum_k \begin{bmatrix} n \\ k \end{bmatrix} \tilde{b}_k = n! 2^{n-1} \qquad (n \geq 1).$$

(c) Let $\{a_n\}$ be the numbers of derangements (= fixed point free permutations) of n letters, and let $\{b_n\}$ be defined as in part (a). Show that

$$\{b_n\} \overset{egf}{\longleftrightarrow} \frac{1-x}{1 + \log(1-x)}.$$

(d) Repeat parts (a)-(c) on the Stirling numbers of the second kind, and discover a few identities of your own that involve them.

(e) Generalize parts (a)-(e) to exponential families.

13. Prove that

$$\sum_k (-1)^{n-k} \binom{2n}{k}^2 = \binom{2n}{n}$$

by exhibiting this sum as a special case of a sum with two free parameters, and by using Snake Oil on the latter.

14. To do a sum that is of the form

$$S(n) = \sum_k f(k)g(n-k),$$

the natural method is to recognize $S(n)$ as $[x^n]\{F(x)G(x)\}$, where F and G are the opsgf's of $\{f_n\}$ and $\{g_n\}$. Use this method to evaluate

$$S(n) = \sum_k \frac{1}{k+1}\binom{2k}{k}\frac{1}{n-k+1}\binom{2n-2k}{n-k}.$$

15.

(a) Prove the following generalization of (4.2.19), and show that it is indeed a generalization. For all $m, n, q \geq 0$, we have

$$\sum_r \binom{m}{r}\binom{n-r}{n-r-q}(t-1)^r = \sum_r \binom{m}{r}\binom{n-m}{n-r-q}t^r.$$

(b) The *Jacobi polynomials* may be defined, for $n \geq 0$, by

$$P_n^{(a,b)}(x) = \sum_k \binom{n+a}{k}\binom{n+b}{n-k}\left(\frac{x-1}{2}\right)^{n-k}\left(\frac{x+1}{2}\right)^k.$$

Use the result of part (a) to show also that

$$P_n^{(a,b)}(x) = \sum_j \binom{n+a+b+j}{j}\binom{n+a}{a}\left(\frac{x-1}{2}\right)^j.$$

(c) Use the result of part (b) and a dash of Snake Oil to show that

$$P_n^{(a,b)}(x) = 2^{-n}(x-1)^{-a}\left[t^{n+a+b}\right]\left\{\frac{(1+x-2t)^{n+a}}{(1-t)^{n+1}}\right\}.$$

16. Prove the following generalization of the sum in example 2: if two sequences $\{f_n\}$ and $\{c_k\}$ are connected by the equations

$$f_n = \sum_k \binom{n+k}{m+2k}c_k \qquad (n \geq 0),$$

where $m \geq 0$ is fixed, then their opsgf's are connected by

$$F(x) = \frac{x^m}{(1-x)^{m+1}}C\left(\frac{x}{(1-x)^2}\right).$$

Say exactly what was special about the sequence $\{c_k\}$ that was used in example 2 that made the result turn out to be so neat in that case.

17. The purpose of this problem is to show the similarity of the method of [WZ] to some well known continuous phenomena.

(a) Let $F(x,y)$, $G(x,y)$ be differentiable functions that satisfy the conditions that $F_x = G_y$ and $\lim_{y\to\pm\infty} G(x,y) = 0$, for all x in a certain interval $a < x < b$. Show that we have the 'identity'

$$\int_{-\infty}^{\infty} F(x,y)dy = const. \qquad (a < x < b).$$

(b) Show, using the result of part (a), that if $f(z)$ is analytic in the strip $-\infty < a < \Re z < b < \infty$, and if $f \to 0$ on all vertical lines in that strip, then the conclusion of part (a) holds, where $F(x,y)$ is the real part of $f(z)$.

(c) Show that the result stated in part (b) is true without using the result of part (a), but using instead the Cauchy integral theorem applied to a suitable rectangle.

(d) Apply these results to $f(z) = e^{z^2}$ and thereby discover the 'identity'

$$\int_{-\infty}^{\infty} e^{-y^2} \cos(2xy)dy = ce^{-x^2} \qquad (x \text{ real}),$$

which states that the function e^{-y^2} is its own Fourier transform. Find c.

18. This problem gives a neat proof of Cayley's formula for the number of trees of n vertices, by showing more, namely that there is a pretty formula for the number of such trees even if the degrees of all vertices are specified.

(a) Let $d_1, \ldots, d_n$ be positive integers whose sum is $2n - 2$. Show that the number of vertex-labeled trees T of n vertices, in which for all $i = 1, \ldots, n$ it is true that d_i is the degree of vertex i of T, is exactly

$$f_n(d_1, \ldots, d_n) = \frac{(n-2)!}{(d_1 - 1)!(d_2 - 1)! \cdots (d_n - 1)!}.$$

(Do this by induction on n. Show that one of the d_i's, at least, must be $=1$, and go from there.)

(b) Find the generating function

$$F_n(x_1, \ldots, x_n) = \sum_{\substack{d_1 + \cdots + d_n = 2n-2 \\ d_1, \ldots, d_n \geq 1}} f_n(d_1, \ldots, d_n)x_1^{d_1} \cdots x_n^{d_n}$$

in a pleasant, explicit form, involving no summation signs.

(c) Let $x_1 = x_2 = \cdots = x_n = 1$ in your answer to part (b), and thereby prove Cayley's result that there are exactly n^{n-2} labeled trees of n vertices.

(d) Use the sieve method to show that if e_k is the number of vertex labeled trees on n vertices of which k are endpoints (vertices of degree 1), then

$$\sum_k e_k x^k = \sum_r \binom{n}{r} r^{n-2}(x-1)^{n-r}.$$

(e) Show that the average number of endpoints that trees of n vertices have is

$$n\left(1 - \frac{1}{n}\right)^{n-2} \sim \frac{n}{e} \qquad (n \to \infty),$$

i.e., *the probability that a random vertex of a tree is an endpoint is about $1/e$.*

19.

(a) If $N(\subseteq S)$ is the number of objects whose set of properties is contained in S, then for all sets T, the number of objects whose set of properties is *precisely* T is

$$N(= T) = \sum_{S \subseteq T} (-1)^{|S|-|T|} N(\subseteq S).$$

(b) Let S be a fixed set of positive integers, and let $h_n(S)$ be the number of hands of weight n, in a certain labeled exponential family, whose card sizes all belong to S. Find the egf of $\{h_n(S)\}$.

(c) Multiply by $(-1)^{|S|-|T|}$ and sum over $S \subseteq T$ to find the egf of $\{\psi_n(T)\}_{n\geq 0}$, the number of hands whose set of distinct card sizes is *exactly* T, in the form (the d's are the deck sizes)

$$\sum_{n\geq 0} \frac{\psi_n(T)}{n!} x^n = \prod_{t\in T}\left(e^{\frac{d_t x^t}{t!}} - 1\right).$$

(d) Let $\rho(n,k)$ be the number of hands of weight n that have exactly k different *sizes* of cards (however many cards they might have!). Sum the result of (c) over all $|T| = k$, etc., to find that

$$\sum_{n,k\geq 0} \frac{\rho(n,k)}{n!} x^n y^k = \prod_{t\geq 1}\left\{1 + y\left(e^{d_t x^t/t!} - 1\right)\right\}.$$

(e) Let c_n be the average number of different sizes of cycles that occur in permutations of n letters. Show that the opsgf of $\{c_n\}$ is

$$\frac{1}{1-x}\sum_{t\geq 1}\left(1 - e^{-x^t/t}\right),$$

and find an explicit formula for c_n.

20. Begin with the set $\{1, 2, \ldots, n\}$. Toss a coin n times, once for each member of the set. Keep the elements that scored 'Heads' and discard the elements that got 'Tails'. You now have a certain subset S of the original set. Call this whole process a 'step'. Now take a step from S. That is, toss a coin for each element of S, and keep those that get 'Heads', getting a sub-subset S', etc. The game halts when the empty set is reached. Let $f(n,k,r)$ be the probability that after k steps, exactly r objects remain.

(a) Find a recurrence relation for f, find the generating function for f, and find f itself.

(b) What is the *average* number of steps in a complete game?

(c) What is the standard deviation of the number of steps in the game?

21. As in section 4.7, let $f(n, k, t)$ be the number of HC-polyominoes that have n cells, in k layers, the highest layer consisting of exactly t cells. Show that the 'grand' three-variable generating function is

$$\sum_{n,k,t} f(n, k, t) x^n y^k z^t = \frac{xyz(1-x)^2((1-xz)(1-x)^2 + x^2y(z-1))}{(1-xz)^2((1-x)^4 - xy(1-x-x^2+x^3+x^2y))}.$$

Note that you do not have to start from scratch here, but instead you can use the results of section 4.7.

22. Let $(a_1, b_1), \ldots, (a_k, b_k)$ be an exact covering sequence. Show that

$$\sum_{j:a_j \text{ is even}} \frac{1}{b_j} = 1/2 = \sum_{j:a_j \text{ is odd}} \frac{1}{b_j}.$$

Generalize this result to other residue classes for the a_j.

23. What is the probability that a random permutation has equal numbers of r-cycles and s-cycles? Express your answer in terms of Bessel functions (see chapter 2). Make a table of your answer, as a function of r and s, for $1 < r < s \le 6$.

24. Find a three term recurrence relation, whose coefficients are polynomials in n, that is satisfied by the quantity shown in (4.8.1), which is the number of convex polyominoes of perimeter $2n + 8$.

25. For $(a_i, b_i)|_{i=1}^{k}$ to be an exact covering sequence it is necessary and sufficient that for all n such that $0 \le n \le N$, n is congruent to a_i mod b_i for exactly one i, where N is the least common multiple of $b_1, \ldots, b_k$.

26.

(a) Develop the generalization of the exponential formula that we were really using in section 4.7. Precisely, suppose that for each $i = 1, 2, 3, \ldots$ we are given a set S_i of positive integers. Let $h(n)$ be the number of hands of weight n that can be formed from a given collection of decks if our choices of cards are restricted by the condition that for each $i = 1, 2, 3, \ldots$, the number of cards of weight i that are chosen for the hand must lie in the set S_i. Then show that

$$\sum_{n \ge 0} h(n) \frac{t^n}{n!} = \prod_{i=1}^{\infty} \exp_{S_i} \left(\frac{d_i t^i}{i!} \right),$$

where $\exp_S(x)$ is the subseries of the exponential series whose indices lie in the set S and, as in chapter 3, d_i is the number of cards in the ith deck.

(b) Find the egf of $\{f(n)\}$, where $f(n)$ is the number of partitions of the
     set $[n]$ in which the number of classes of size 2 is divisible by 2 and
     the number of classes of size 3 is divisible by 3, etc.

27. In order that $(a_i, b_i)|_{i=1}^k$ be an exact covering sequence of residues and
moduli, it is necessary and sufficient that [Fr]

$$\sum_{i=1}^{k} b_i^{n-1} B_n\left(\frac{a_i}{b_i}\right) = B_n \qquad (n = 0, 1, 2, \ldots)$$

where the $\{B_n\}$ are the Bernoulli numbers (defined by (2.5.8)), and the
$B_n(x)$ are the *Bernoulli polynomials*, defined by

$$\frac{te^{xt}}{e^t - 1} = \sum_{n=0}^{\infty} \frac{B_n(x)t^n}{n!}.$$

28. Find a formula for the number of square roots that a permutation has.
What kind of a permutation has a unique square root?

Chapter 5
Analytic and asymptotic methods

In the preceding chapters we have emphasized the formal aspects of the theory of generating functions, as opposed to the analytic theory. One of the attractions of the subject, however, is how easily one can shift gears from thinking of generating functions as mere clotheslines for sequences to regarding them as analytic functions of complex variables. In the latter state of mind, one can deduce many properties of the sequences that are generated that would be inaccessible to purely formal approaches. Notable among these properties are the asymptotic growth rates of the sequences, which are probably the main focus of the analytic side of the theory. Hence in this chapter we will develop some of the analytic machinery that is invaluable for such studies. For an introduction to asymptotics and definitions of all of the symbols of asymptotics, see, for example, chapter 4 of [Wil].

5.1 The Lagrange Inversion Formula

The Lagrange Inversion Formula is a remarkable tool for solving certain kinds of functional equations, and at its best it can give explicit formulas where other approaches run into stone walls. The form of the functional equation that the LIF can help with is

$$u = t\phi(u). \tag{5.1.1}$$

Here ϕ is a given function of u, and we are thinking of the equation as determining u as a function of t. That is, we are 'solving for u in terms of t.'

We found one example of such an equation in section 3.12. There we saw that if $T(x)$ is the egf for the numbers of rooted labeled trees of each number of vertices, then $T(x)$ satisfies the equation $T = xe^T$, which is indeed of the form (5.1.1), with $\phi(u) = e^u$.

The general problem is this: suppose we are given the power series expansion of the function $\phi = \phi(u)$, convergent in some neighborhood of the origin (of the u-plane). How can we find the power series expansion of the solution of (5.1.1), $u = u(t)$, in some neighborhood of the origin (in the t-plane)? The answer is surprisingly explicit, and it even allows us to find the expansion of some function of the solution $u(t)$.

Theorem 5.1.1. *(The Lagrange Inversion Formula) Let $f(u)$ and $\phi(u)$ be formal power series in u, with $\phi(0) = 1$. Then there is a unique formal power series $u = u(t)$ that satisfies (5.1.1). Further, the value $f(u(t))$ of f at that root $u = u(t)$, when expanded in a power series in t about $t = 0$, satisfies*

$$[t^n]\{f(u(t))\} = \frac{1}{n}[u^{n-1}]\{f'(u)\phi(u)^n\}. \tag{5.1.2}$$

Proof.   First we note that it suffices to prove the theorem in the case where f and ϕ are polynomials. Indeed, if n is fixed, and if f and ϕ are full formal power series, then suppose that we truncate both of those series by discarding all terms that involve powers u^k for $k > n$. If the result is true for these polynomials then it remains true for the original untruncated series, since the higher order terms that were discarded do not affect (5.1.2), for the fixed n, at all.

Therefore we now suppose that f and ϕ are polynomials.

We will first make a formal computation, and then discuss the range of validity of the results. We have

$$
\begin{aligned}
\left[u^{n-1}\right]\left\{f'(u)\phi(u)^n\right\} &= \left[u^{n-1}\right]\left\{f'(u)(u/t)^n\right\}\\
&= \left[u^{-1}\right]\left\{f'(u)/t^n\right\}\\
&= \frac{1}{2\pi i}\int \frac{f'(u)}{t(u)^n}du\\
&= \frac{1}{2\pi i}\int t^{-n}f'(u(t))u'(t)dt \qquad (5.1.3)\\
&= \left[t^n\right]\left\{t(d/dt)f(u(t))\right\}\\
&= n\left[t^n\right]f(u(t)).
\end{aligned}
$$

In the above, the first equality comes from (5.1.1), the second is trivial, and the third is the residue theorem of complex integration, in which the integrand is a function of u and the contour is, say, a small circle enclosing 0.

The fourth equality needs some discussion. Consider the behavior of the function $g(u) = u/\phi(u)$ near the origin. Since $\phi(0) = 1$, the function ϕ remains nonzero in a neighborhood of 0. Hence g is analytic there, and it has a power series development of the form $u + cu^2 + \cdots$. It follows that g is a 1-1 conformal map near 0. Hence it has a well defined inverse mapping that is itself analytic near 0. Thus (5.1.1) has a unique solution $u = u(t)$ in some neighborhood $\Re$ of $t = 0$, and u is an analytic function of t there. If the contour of integration in the integral that appears after the fourth equals sign above is a circle around $t = 0$ that lies in $\Re$, then the sign of equality is valid simply as a change of variable from u to t in the integral.

The remaining equalities are trivialities, and the proof is complete. ∎

Example 1.

In section 3.12 we embarked on proving theorem 3.12.1, to the effect that there are exactly n^{n-2} labeled trees of n vertices. We found that if $\mathcal{D}(x) \overset{egf}{\longleftrightarrow} \{t_n\}$, where t_n is the number of *rooted* labeled trees of n vertices, then $\mathcal{D}(x)$ satisfies the functional equation

$$
\mathcal{D}(x) = xe^{\mathcal{D}(x)}. \qquad (5.1.4)
$$

Now, with the Lagrange Inversion Formula, we can actually solve (5.1.4), because it is the case $\phi(u) = e^u$ of (5.1.1). If we take $f(u) = u$, then according to (5.1.2),

$$[x^n]\,\mathcal{D}(x) = (1/n)\left[u^{n-1}\right]\{\phi(u)^n\}$$
$$= (1/n)\left[u^{n-1}\right]\{e^{nu}\}$$
$$= (1/n)\frac{n^{n-1}}{(n-1)!}$$
$$= \frac{n^{n-1}}{n!}.$$

Hence,

$$\frac{t_n}{n!} = \frac{n^{n-1}}{n!},$$

and $t_n = n^{n-1}$. But t_n is the number of rooted trees of n vertices, and every labeled tree contributes n rooted labeled trees, so the number of labeled trees of n vertices is n^{n-2}, which completes the proof of theorem 3.12.1. ∎

Example 2.

At the end of chapter 4 we discussed inverse pairs of summation formulas and gave some examples beyond the Möbius inversion formula. Here we'll derive a much fancier example with the help of the LIF. We propose to show that if two sequences $\{a_n\}$ and $\{b_n\}$ are related by

$$b_n = \sum_k \binom{k}{n-k} a_k, \qquad (5.1.5)$$

then we have the inversion

$$na_n = \sum_k \binom{2n-k-1}{n-k}(-1)^{n-k}kb_k. \qquad (5.1.6)$$

Indeed, if (5.1.5) holds, then, as we did so often in section 4.3, multiply by x^n, sum on n, and interchange the k and n summations, to get

$$B(x) = \sum_k a_k x^k \sum_n \binom{k}{n-k}x^{n-k}$$
$$= \sum_k a_k x^k (1+x)^k \qquad (5.1.7)$$
$$= A(x(1+x)),$$

where A, B are the opsgf's of the sequences. Now we can solve for A in terms of B by setting $y = x + x^2$. Then if we read (5.1.7) backwards, we find that

$$A(y) = B(x(y)), \qquad (5.1.8)$$

where $x(y)$ is the solution of the quadratic equation $y = x + x^2$ that vanishes when $y = 0$.

Now we *could*, of course, solve the quadratic explicitly. If we were to do that (which we won't) we would find from (5.1.8) that

$$A(y) = B\left(\frac{\sqrt{1 + 4y} - 1}{2}\right),$$

and then we would want to find a nice formula for the coefficient of y^n on the right hand side for every n. That, in turn, would require a nice formula for

$$[y^n]\left\{\frac{\sqrt{1 + 4y} - 1}{2}\right\}^k \tag{5.1.9}$$

for every n and k. Instead of trying to deal with (5.1.9) by explicitly raising the quantity in braces to the kth power and working with the square root, it is a lot more elegant to let the LIF do the hard work.

We begin by rephrasing the question (5.1.9) *implicitly*, rather than explicitly. What we want is

$$[y^n]\{x(y)^k\},$$

where $x = x(y)$ is the solution of $y = x + x^2$ that is 0 at $y = 0$. In terms of the LIF, we write the equation as

$$x = \frac{y}{1 + x},$$

which is of the form (5.1.1) with $\phi(u) = 1/(1 + u)$. Further, since we want the coefficients of the kth power of $x(y)$, the function $f(u)$ in the LIF is now $f(u) = u^k$.

The conclusion (5.1.2) of the LIF tells us that

$$\begin{aligned}
[y^n]\{x(y)\}^k &= (1/n)[x^{n-1}]\left\{\frac{kx^{k-1}}{(1 + x)^n}\right\} \\
&= (k/n)[x^{n-k}]\frac{1}{(1 + x)^n} \\
&= (k/n)(-1)^{n-k}\binom{2n - k - 1}{n - k},
\end{aligned}$$

and we are all finished with the proof of (5.1.6).

It should be particularly noted that the LIF is as adept at computing the coefficients of the kth power of the unknown function as those of the unknown function itself. The function f in the statement of the LIF *simply specifies the function of the unknown function whose coefficients we would like to know*. The LIF then hands us those coefficients.

5.2 Analyticity and asymptotics (I): Poles

Suppose we have found the generating function $f(z)$ for a certain sequence of combinatorial numbers that interests us. Next we might want to find the asymptotic behavior of the sequence, i.e., to find a simple function of n that affords a good approximation to the values of our sequence when n is large.

The first law of doing asymptotics is: *look for the singularity or singularities of $f(z)$ that are nearest to the origin.* The reason is that $f(z)$ is analytic precisely in the largest circle centered at the origin that contains no singularities, and we find the radius of that circle by finding the singularities nearest to the origin. Once we have that radius, we have also the radius of convergence of the power series $f(z)$. Once we have the radius of convergence we know something about the sizes of the coefficients when n is large, as in theorem 2.4.3. By various refinements of this process we can discover more detailed information.

Therefore, in this and the following sections we will study the influence of the singularities of analytic functions on the asymptotic behavior of their coefficients. These methods, taken together, provide a powerful technique for obtaining the asymptotics of combinatorial sequences, and provide yet another justification, if one were needed, of the generating function approach.

In this section we will concentrate on functions whose only singularities are poles.

Let $f(z)$ be analytic in some region of the complex plane that includes the origin, with the exception of a finite number of singularities. If R is the smallest of the moduli of these singularities, then f is analytic in the disk $|z| < R$, so this will be precisely the disk in which its power series expansion about the origin converges.

Conversely, if the power series expansion of a certain generating function f converges in the disk $|z| < R$ but in no larger disk centered at the origin, then there are one or more singularities of the function f on the circumference $|z| = R$.

A number of methods for dealing with questions of asymptotic growth of coefficient sequences rely on the following strategy: find a simple function g that has the same singularities that f has on the circle $|z| = R$. Then $f - g$ is analytic in some larger disk, of radius $R' > R$, say.

Then, according to theorem 2.4.3 (q.v.), the power series coefficients of $f - g$ will be $< (\frac{1}{R'} + \epsilon)^n$ for large n, and therefore they will be much smaller than the coefficients of f itself. The latter, according to the same theorem 2.4.3, will infinitely often be as large as $(\frac{1}{R} - \epsilon)^n$.

Therefore we will be able to find the most important aspects of the growth of the coefficients of f by looking at the growth of the coefficients of g.

The strategy that wins, therefore, is that of finding a simple function that mimics the singularities of the function that one is interested in, and then of using the growth of the coefficients of the simple function for the estimate.

These considerations come through most clearly in the case of a *meromorphic* function $f(z)$, i.e., one that is analytic in the region with the exception of a finite number of *poles*, and so we will study such functions first. The idea is that near a pole z_0, a meromorphic function is well approximated by the *principal part* of its Laurent expansion, i.e., by the finite number of terms of the series that contain $(z - z_0)$ raised to negative powers.

Example 1.

The function $f(z) = e^z/(z-1)$ is meromorphic in the whole finite plane. Its only singularity is at $z_0 = 1$, and the principal part at that singularity is $e/(z-1)$. Hence the function $f(z) - e/(z-1)$ is analytic in the whole plane. Thus if $\{c_n\}$ are the power series coefficients of f about the origin, and $\{d_n\}$ are the same for the function $e/(z-1)$, the difference $c_n - d_n$ is small when n is large. In fact, theorem 2.4.3 guarantees that for every $\epsilon > 0$ we have $|c_n - d_n| < \epsilon^n$ for all large enough n. Therefore the 'unknown' coefficients of $f(z)$ are very well approximated by those of the simple function $e/(z-1)$.

It is easy to work out this particular example completely to see just how the machine works. The expansion of $e^z/(z-1)$ about the origin is (see Rule 5 of section 2.2)

$$\frac{e^z}{(z-1)} = -\sum_{n \geq 0}\{1 + 1 + 1/2! + \cdots + 1/n!\}z^n.$$

On the other hand, the expansion of $e/(z-1)$ is

$$\frac{e}{(z-1)} = -e - ez - ez^2 - ez^3 - \cdots.$$

Hence the act of replacing the function by its principal part yields in one step the approximation of the true coefficient of z^n, which is the nth partial sum of the power series for $-e$, by $-e$ itself, which is a smashingly good approximation indeed. The reason for the great success in this case is that merely subtracting off one principal part from the function expands its disk of analyticity from radius $=1$ to radius $= \infty$.

Here's another way to look at this example, without mentioning any of the heavy machinery. Consider the innocent fact that

$$f(z) = \frac{e^z}{z-1} = \frac{e}{z-1} + \frac{e^z - e}{z - 1}.$$

In the first term, the power series coefficients are all equal to $-e$. The second term has no singularities at all in the finite plane, i.e., it is an *entire function* of z. By theorem 2.4.3, its coefficients are $O(\epsilon^n)$ for every positive ϵ. Therefore the coefficients of $f(z)$ are

$$= -e + O(\epsilon^n) \qquad (n \to \infty)$$

for every $\epsilon > 0$. ∎

In less favorable cases one may have to subtract off several principal parts in order to increase the size of the disk of analyticity at all (i.e., if there are several poles on the circumference), and even then it may increase only a little bit if there are other singularities on a slightly larger circle.

If f is meromorphic in $\Re$, let z_0 be a pole of f of order r, $1 \le r < \infty$. Then in some punctured disk centered at z_0, f has an expansion

$$f(z) = \sum_{j=1}^{r} \frac{a_{-j}}{(z - z_0)^j} + \sum_{j=0}^{\infty} a_j (z - z_0)^j. \qquad (5.2.1)$$

The first one of the two sums above, the one containing the negative powers of $(z - z_0)$, is called the *principal part* of the expansion of f around the singularity z_0, and we will denote it by $PP(f; z_0)$. The function $f - PP(f; z_0)$ is analytic at z_0. That is to say, we can remove the singularity by subtracting off the principal part.

If, besides z_0, there are other poles of f on the same circle $|z| = R = |z_0|$, then let $z_1, \ldots, z_s$ be all such poles. The function

$$h(z) = f(z) - PP(f; z_0) - PP(f; z_1) - \cdots - PP(f; z_s) \qquad (5.2.2)$$

is regular (analytic) at every one of the points $\{z_j\}_0^s$. But f had no other singularities on that circle, so h is analytic in a circle centered at the origin that has radius R', where $R' > R$.

That means, by theorem 2.4.3 again, that the power series coefficients of h, about the origin, cannot grow faster than

$$\left(\frac{1}{R'} + \epsilon \right)^n$$

for all large n. Thus, if $f \overset{ops}{\longrightarrow} \{a_n\}$, and if

$$g(z) = PP(f; z_0) + PP(f; z_1) + \cdots + PP(f; z_s) \overset{ops}{\longrightarrow} \{b_n\}, \qquad (5.2.3)$$

then

$$a_n = b_n + O\left(\left(\frac{1}{R'} + \epsilon \right)^n \right) \qquad (n \to \infty).$$

We may then be well on our way towards finding the asymptotic behavior of the coefficients of f.

Indeed, let us now study the power series coefficients, about the origin, of the sum of the principal parts that are shown in (5.2.3). We have, if z_0 is a pole of multiplicity r,

$$
\begin{aligned}
PP(f; z_0) &= \sum_{j=1}^{r} \frac{a_{-j}}{(z - z_0)^j} \\
&= \sum_{j=1}^{r} \frac{(-1)^j a_{-j}}{z_0^j (1 - (z/z_0))^j} \\
&= \sum_{j=1}^{r} \frac{(-1)^j a_{-j}}{z_0^j} \sum_{n \geq 0} \binom{n+j-1}{n} (z/z_0)^n \\
&= \sum_{n \geq 0} z^n \left\{ \sum_{j=1}^{r} \frac{(-1)^j a_{-j}}{z_0^{n+j}} \binom{n+j-1}{j-1} \right\}.
\end{aligned}
\tag{5.2.4}
$$

We see, therefore, that a pole of order r at z_0, of a function f, contributes

$$
\sum_{j=1}^{r} \frac{(-1)^j a_{-j}}{z_0^{n+j}} \binom{n+j-1}{j-1}
\tag{5.2.5}
$$

to the coefficient of z^n in f.

The basic theorem, which asserts that we can well approximate the coefficients of a meromorphic function by the coefficients of the principal parts at its poles of smallest modulus, can be stated as follows:

Theorem 5.2.1. *Let f be analytic in a region $\Re$ containing the origin, except for finitely many poles. Let $R > 0$ be the modulus of the pole(s) of smallest modulus, and let $z_0, \ldots, z_s$ be all of the poles of $f(z)$ whose modulus is R. Further, let $R' > R$ be the modulus of the pole(s) of next-smallest modulus of f, and let $\epsilon > 0$ be given. Then*

$$
[z^n] f(z) = [z^n] \left\{ \sum_{j=0}^{s} PP(f; z_j) \right\} + O\left(\left(\frac{1}{R'} + \epsilon \right)^n \right).
\tag{5.2.6}
$$

Proof. By theorem 2.4.3, this theorem will be proved as soon as we establish that if we subtract from $f(z)$ the sum of all of its principal parts from singularities on the circle $|z| = R$, then the resulting function is analytic in the larger disk $|z| < R'$. Consider the moment when we subtract $PP(f; z_0)$ from $f(z)$. Certainly the resulting function, g, say, is analytic at z_0. Next, however, we subtract $PP(f; z_1)$ from g instead of subtracting $PP(g; z_1)$ from g. We claim that this doesn't matter, i.e., that

$$
PP(f - PP(f; z_0); z_1) = PP(f; z_1).
$$

To see this, observe that

$$PP(f - PP(f; z_0); z_1) = PP(f; z_1) - PP(PP(f; z_0); z_1).$$

But the second term on the right vanishes because $PP(f; z_0)$ is analytic at z_1.

By induction on s, the result follows. ∎

Example 1. Ordered Bell numbers

We now investigate the asymptotic behavior of the 'ordered Bell numbers.' These are defined as follows: a set of n elements has $\left\{ {n \atop k} \right\}$ partitions into k classes. If we now regard the order of the classes as important, but not the order of the elements within the classes, then we see that $[n]$ has $k! \left\{ {n \atop k} \right\}$ *ordered partitions into k classes.* The ordered Bell number $b(n)$ is the total number of ordered partitions of $[n]$, i.e., it is $\sum_k k! \left\{ {n \atop k} \right\}$.

Our question concerns the growth of $\{\tilde{b}(n)\}$ when n is large. To find a nice formula for these numbers, multiply both sides of the identity (4.2.16) by e^{-y} and integrate from 0 to ∞. This gives the neat result that

$$\tilde{b}(n) = \sum_{r \geq 0} \frac{r^n}{2^{r+1}}. \tag{5.2.6}$$

Then the exponential generating function of the ordered Bell numbers is*

$$f(z) = \sum_{n \geq 0} \frac{\tilde{b}(n)}{n!} z^n = \frac{1}{2 - e^z}. \tag{5.2.7}$$

We're in luck! The generating function $f(z)$ has only simple poles, namely at the points $\log 2 \pm 2k\pi i$ for all integer k. The principal part at the pole $z_0 = \log 2$ is $(-1/2)/(z - \log 2)$. That principal part all by itself contributes

$$\frac{1}{2(\log 2)^{n+1}}$$

to the coefficient of z^n. There are no other singularities of $f(z)$ on the circle of radius $\log 2$ centered at the origin. Hence

$$h(z) = f(z) - \frac{(-1/2)}{(z - \log 2)}$$

is analytic in the larger circle that extends from the origin to $\log 2 + 2\pi i$. The radius of that circle is

$$\rho = \sqrt{(\log 2)^2 + 4\pi^2} = 6.321 \ldots.$$

---

* Be sure to work this out for yourself.

Hence the coefficients of $h(z)$ are $O((.16)^n)$. Altogether, we have shown that *the ordered Bell numbers $\tilde{b}(n)$ are of the form*

$$\tilde{b}(n) = \frac{1}{2(\log 2)^{n+1}} n! + O((.16)^n n!), \qquad (5.2.8)$$

which is not bad for so little effort invested. More terms of the asymptotic expansion can be produced as desired from the principal parts of $f(z)$ at its remaining poles, taken in nondescending order of their absolute values. The reader should look into the contribution of the next two poles together, which are complex conjugates of each other.

Below we show a table of some values of n, $\tilde{b}(n)$, and $n!/(2(\log 2)^{n+1})$.

n	1	2	3	5	10
$\tilde{b}(n)$	1	3	13	541	102247563
$n!/(2(\log 2)^{n+1})$	1.04	3.002	12.997	541.002	102247563

The agreement is astonishingly close. Basically all we have done is to use the Taylor coefficients of the series for $1/(2(\log 2 - z))$ as approximations to the coefficients of the series for $1/(2 - e^z)$. Yet we are rewarded with a superb approximation. ∎

Example 2. Permutations with no small cycles

Fix a positive integer q. Let $f(n,q)$ be the number of permutations of n letters whose cycles all have lengths $> q$. We want the asymptotic behavior of $f(n,q)$.

By exercise 11 of chapter 3, the egf of $\{f(n,q)\}_0^\infty$ is

$$f_q(z) = \exp \sum_{n>q} \frac{z^n}{n}$$

$$= \exp \left\{ \log \frac{1}{1-z} - \sum_{1 \le n \le q} \frac{z^n}{n} \right\} \qquad (5.2.9)$$

$$= \frac{1}{1-z} e^{-\{z+\cdots+z^q/q\}}.$$

The only singularity of $f_q(z)$ in the finite plane is a pole of order 1 at $z = 1$ with principal part $e^{-H_q}/(1-z)$, where

$$H_q = 1 + \frac{1}{2} + \frac{1}{3} + \cdots + \frac{1}{q}$$

is the qth harmonic number.

This is the kind of situation where we get very accurate asymptotic estimates, because the difference between the function and its principal part at $z = 1$ is

$$h(z) = f_q(z) - \frac{e^{-H_q}}{1 - z}$$

$$= \frac{e^{-\{z + \cdots + z^q/q\}} - e^{-H_q}}{1 - z},$$

and is analytic in the whole plane, i.e., is an entire function. Again, by theorem 2.4.3, the nth coefficient of $h(z)$ is $O(\epsilon^n)$ as $n \to \infty$ for every $\epsilon > 0$. Therefore,

$$\frac{f(n, q)}{n!} = e^{-H_q} + O(\epsilon^n) \qquad (n \to \infty). \tag{5.2.10}$$

The strikingly small error term in this estimate suggests that for each fixed q the probability that an n-permutation has no cycles of length $\leq q$ should be very nearly independent of n. Consider the case $q = 1$ to get some of the flavor of what is going on here. Then $f(n, 1)/n!$ is the probability that an n-permutation has no fixed point. But we saw in (4.2.10) that

$$f(n, 1)/n! = e^{-1}_{|n} = 1 - 1 + 1/2 - \cdots + (-1)^n/n!$$

$$= e^{-1} + O(1/n!).$$

Indeed, the probability is very nearly independent of n, and the error involved in using the principal part is $O(\epsilon^n)$ for every positive ϵ. ∎

The two examples above have shown the method at work in situations where it was atypically accurate. More commonly one finds not just one pole of order 1 in the entire plane, but many poles of various multiplicities. The method remains the same in such cases, but a lot more work may be necessary in order to get estimates of reasonable accuracy. An example that shows this kind of phenomenon was worked out in section 3.15, in connection with the money-changing problem. In fact, the proof of Schur's theorem (Theorem 3.15.2) was an exercise in the use of principal parts at the poles of a meromorphic function. The importance of the single dominant singularity was, in that case, much less, though it was enough to get the theorem proved!

5.3 Analyticity and asymptotics (II): Algebraic singularities

Again, let $f(z)$ be analytic in some region that contains the origin, but now suppose that the singularity z_0 of f that is nearest to 0 is not a pole, but is an algebraic singularity (*branch point*). What that means is that $f(z) = (z_0 - z)^\alpha g(z)$, where g is analytic at z_0 and α is not an integer, but is a real number.

A case in point was given in (3.9.1), where we found that the egf for the numbers of graphs of n vertices whose vertex-degrees are all equal to 2 is

$$f(z) = \frac{e^{-z/2-z^2/4}}{\sqrt{1-z}}. \tag{5.3.1}$$

In this section we will derive the theorem ('Darboux's lemma') that allows us to deduce the asymptotics of sequences with this kind of a generating function. What it all boils down to is that one should do exactly the same thing in this case as in the case of meromorphic functions, and the right answer will fall out. The proof that this is indeed valid, however, is more demanding in the present case. We follow the proof in [KnW].

By considering $f(zz_0)$ instead of $f(z)$, if necessary, we see that we can assume without loss of generality that $z_0 = 1$. Hence we are dealing with a function f that is analytic in the unit disk, and which has a branch point at $z = 1$. We will also assume, until further notice, that $z_0 = 1$ is the *only* singularity that f has in some disk $|z| < 1 + \eta$, where $\eta > 0$.

After the lessons of the previous section on meromorphic functions, here's how we might proceed in this case. First we have $f(z) = (1-z)^\alpha g(z)$, where g is analytic at $z = 1$. That being the case, we can expand g in a power series

$$g(z) = \sum_{k \geq 0} g_k (1-z)^k$$

that converges in a neighborhood of $z = 1$. Hence f itself has an expansion

$$f(z) = \sum_{k \geq 0} g_k (1-z)^{k+\alpha}. \tag{5.3.2}$$

By analogy with the procedure for meromorphic functions, we might expect that each successive term in the above series expansion generates the next term of the asymptotic expansion of the coefficients of f. That is in fact true. The dominant behavior of the coefficient of z^n in $f(z)$ comes from the first term in (5.3.2). That is, the simple function $g_0(1-z)^\alpha$ has, for its coefficient of z^n, the main contribution to that coefficient of f, etc.

We will now prove all of these things.

Lemma 5.3.1. *Let $\{a_n\}$, $\{b_n\}$ be two sequences that satisfy (a) $a_n = O(n^{-\gamma})$ and (b) $b_n = O(\theta^n)$ $(0 < \theta < 1)$. Then*

$$\sum_k a_k b_{n-k} = O(n^{-\gamma}).$$

Proof. We have first (the C's are not all the same constant)

$$\left| \sum_{0 \leq k \leq n/2} a_k b_{n-k} \right| \leq \left\{ \max_{0 \leq k \leq n/2} |a_k| \right\} \left\{ \sum_{0 \leq k \leq n/2} C\theta^{n-k} \right\}$$

$$\leq \max\{C, Cn^{-\gamma}\}\{C\theta^{n/2}\}$$

$$\leq C\tilde{\theta}^n \qquad (0 < \tilde{\theta} < 1).$$

Further,

$$\left| \sum_{n/2 < k \le n} a_k b_{n-k} \right| \le \left\{ \max_{n/2 < k \le n} |a_k| \right\} \left\{ \sum_{n/2 < k \le n} \theta^{n-k} \right\}$$

$$\le C n^{-\gamma}.$$

Lemma 5.3.2. *If $\beta \notin \{0, 1, 2, \ldots\}$, then*

$$[z^n](1-z)^\beta \sim \frac{n^{-\beta-1}}{\Gamma(-\beta)}. \tag{5.3.3}$$

Proof. We have

$$[z^n](1-z)^\beta = \binom{\beta}{n} (-1)^n$$

$$= \binom{n-\beta-1}{n}$$

$$= \frac{\Gamma(n-\beta)}{\Gamma(-\beta)\Gamma(n+1)},$$

and the result follows from Stirling's formula, which is

$$\Gamma(n+1) = n! \sim \left(\frac{n}{e}\right)^n \sqrt{2\pi n} \qquad (n \to \infty).$$

Lemma 5.3.3. *Let $u(z) = (1-z)^\gamma v(z)$, where $v(z)$ is analytic in some disk $|z| < 1 + \eta$, $(\eta > 0)$. Then*

$$[z^n]u(z) = O(n^{-\gamma-1}). \tag{5.3.4}$$

Proof. Apply lemma 5.3.1 with $a_n = [z^n](1-z)^\gamma$ and $b_n = [z^n]v(z)$. Since v is analytic in a disk $|z| < 1 + \eta$, we have $b_n = O(\theta^n)$. The result follows by lemma 5.3.2. ■

Theorem 5.3.1. *(Darboux) Let $v(z)$ be analytic in some disk $|z| < 1 + \eta$, and suppose that in a neighborhood of $z = 1$ it has the expansion $v(z) = \sum v_j (1-z)^j$. Let $\beta \notin \{0, 1, 2, \ldots\}$. Then*

$$[z^n]\left\{(1-z)^\beta v(z)\right\} = [z^n]\left\{\sum_{j=0}^{m} v_j (1-z)^{\beta+j}\right\} + O(n^{-m-\beta-2})$$

$$= \sum_{j=0}^{m} v_j \binom{n-\beta-j-1}{n} + O(n^{-m-\beta-2}). \tag{5.3.5}$$

Proof. We have

$$(1-z)^{\beta}v(z) - \sum_{j=0}^{m} v_j(1-z)^{\beta+j} = \sum_{j>m} v_j(1-z)^{\beta+j}$$
$$= (1-z)^{\beta+m+1}\tilde{v}(z),$$

where the regions of analyticity of $\tilde{v}$ and of v are the same. The result now follows from lemma 5.3.3. ■

Example 1. 2-regular graphs

For the exponential generating function $f(z)$, in (5.3.1), of the number of 2-regular graphs of n vertices, we have $f(z) = (1-z)^{\beta}v(z)$ with $\beta = -1/2$ and $v(z) = \exp\{-z/2 - z^2/4\}$. The first few terms of the expansion of $v(z)$ about $z = 1$ are

$$e^{-z/2-z^2/4} = e^{-3/4} + e^{-3/4}(1-z) + \frac{1}{4}e^{-3/4}(1-z)^2 + \cdots$$

Then according to (5.3.5) this expansion of $v(z)$ around $z = 1$ 'lifts' to an asymptotic formula for the coefficients of $f(z)$, which are in this case $\gamma(n)/n!$, where $\gamma(n)$ is the number of 2-regular graphs of n vertices. If we use (5.3.5) with $m = 2$, we obtain

$$\frac{\gamma(n)}{n!} = e^{-3/4}\binom{n-1/2}{n} + e^{-3/4}\binom{n-3/2}{n} + \frac{1}{4}e^{-3/4}\binom{n-5/2}{n} \quad (5.3.6)$$
$$+ O(n^{-7/2}).$$

If we like, we can further simplify the answer by using the known asymptotic expansion of the binomial coefficient

$$\binom{n-\alpha-1}{n} \approx \frac{n^{-\alpha-1}}{\Gamma(-\alpha)}\left[1 + \frac{\alpha(\alpha+1)}{2n} + \frac{\alpha(\alpha+1)(\alpha+2)(3\alpha+1)}{24n^2} + \cdots\right]. \quad (5.3.7)$$

If this be substituted into (5.3.6), the result is

$$\gamma(n) \approx \frac{n!e^{-3/4}}{\sqrt{n\pi}}\left\{1 - \frac{5}{8n} + \frac{1}{128n^2} + \cdots\right\}. \quad (5.3.8)$$

The form of Darboux's method that we have proved applies when there is just one algebraic singularity on the circle of convergence. The method can be extended to several such singularities. We quote without proof a more general result of this kind ([Sz], thm. 8.4):

Theorem 5.3.2. *(Szegö) Let $h(w)$ be analytic in $|w| < 1$ and suppose it has a finite number of singularities $\{e^{i\phi_k}\}_1^r$ on $|w| = 1$. Suppose that in the neighborhood of each singularity $e^{i\phi_k}$ there is an expansion*

$$h(w) = \sum_{\nu \geq 0} c_\nu^{(k)} (1 - we^{-i\phi_k})^{\alpha_k + \nu\beta_k},$$

where $\beta_k > 0$. Then the following is a complete asymptotic series for the coefficients of $h(w)$:

$$[w^n]h(w) \approx \sum_{\nu \geq 0} \sum_{k=1}^r c_\nu^{(k)} \binom{\alpha_k + \nu\beta_k}{n} (-e^{i\phi_k})^n.$$

5.4 Analyticity and asymptotics (III): Hayman's method

In the previous two sections we have seen how to handle the asymptotics of sequences whose generating functions have singularities in the finite plane. Essentially, one looks for the singularity(ies) nearest the origin, finds simple functions whose behavior near the singularities is the same as that of the generating function in question, and then proves that the asymptotic behavior of the coefficients of that generating function is the same as that of the coefficients of the simple functions that behave the same way near the singularities.

But what shall we do if the generating function doesn't have any singularities, i.e., if it is an *entire function*?

Example 1. The coefficients of e^z

Consider the function e^z. The coefficient of z^n in e^z is $1/n!$. Can we think of some fairly general method for handling the asymptotics of entire functions, which in this case will derive Stirling's formula for us?

Here's how we might begin. By Cauchy's formula we have

$$\frac{1}{n!} = \frac{1}{2\pi i} \int \frac{e^z dz}{z^{n+1}},$$

where the contour of integration is some simple closed curve that encloses the origin. If we use for the contour a circle of radius r centered at the origin, then by taking absolute values we find that

$$\frac{1}{n!} \leq \frac{1}{2\pi} \max_{|z|=r} \left\{ \frac{|e^z|}{|z|^{n+1}} \right\} (2\pi r)$$

$$= \frac{e^r}{r^n}.$$

Since e^z is an entire function the value of $r > 0$ is entirely up to us, so we might as well choose it to minimize the upper bound that we will obtain. But

$$\min_{r>0} \frac{e^r}{r^n} \tag{5.4.1}$$

is attained at $r = n$, so the best possible estimate that we can get from this argument is that

$$\frac{1}{n!} \leq (\frac{e}{n})^n.$$

If we compare this with Stirling's formula

$$\frac{1}{n!} \sim \frac{1}{\sqrt{2n\pi}}(\frac{e}{n})^n,$$

we see that we haven't done too badly, since our rather crude estimate differs from the 'truth' by only a factor of about $1/\sqrt{2n\pi}$.

To do better than this we are going to have to treat the variation of e^z around the contour of integration with a little more respect, and not just replace it by the maximum absolute value that it attains. Indeed, for most of the way around the circumference $|z| = r$, the absolute value of e^z is considerably smaller than e^r. It is only in a small neighborhood of the point $z = r$ that it is nearly that large.

In his 1956 paper *A generalisation of Stirling's formula*, W. K. Hayman [Ha] developed machinery of considerable power for dealing more precisely with this kind of situation. Further, his method is uncommonly useful for generating functions that arise in combinatorial theory, because these tend to have nonnegative real coefficients. For that reason, on any circle centered at the origin, such a function will be largest in modulus at the positive real point on that circle. Hayman's method is strongest on just such functions.

Hayman's machinery applies not only to entire functions, but to all analytic functions, even those with singularities in the finite plane. In practice it has most often been used on entire functions, mainly because, as we have seen, other methods are available when singularities exist in the finite plane.

Let $f(z)$ be analytic in a disk $|z| < R$ in the complex plane, where $0 < R \leq \infty$. Suppose further that $f(z)$ is an *admissible* function for the method. Operationally, that simply means that $f(z)$ is a function on which the method works. We will give some sufficient conditions for admissibility below.

Define

$$M(r) = \max_{|z|=r}\{|f(z)|\}. \tag{5.4.2}$$

It will be a consequence of the admissibility conditions that

$$M(r) = f(r) \tag{5.4.3}$$

for all large enough r. This is because, as we remarked above, the method is aimed at functions that take their largest values in the direction of the positive real axis.

Next define two auxiliary functions,

$$a(r) = r\frac{f'(r)}{f(r)} \tag{5.4.4}$$

and

$$b(r) = ra'(r) = r\frac{f'(r)}{f(r)} + r^2\frac{f''(r)}{f(r)} - r^2\left(\frac{f'(r)}{f(r)}\right)^2. \tag{5.4.5}$$

The main result is the following:

Theorem 5.4.1. *(Hayman) Let* $f(z) = \sum a_n z^n$ *be an admissible function. Let* r_n *be the positive real root of the equation* $a(r_n) = n$, *for each* $n = 1, 2, \ldots$, *where* $a(r)$ *is given by eq. (5.4.4) above. Then*

$$a_n \sim \frac{f(r_n)}{r_n^n \sqrt{2\pi b(r_n)}} \qquad \text{as } n \to +\infty, \tag{5.4.6}$$

where $b(r)$ *is given by (5.4.5) above.*

It will be noted that the recipe itself is quite straightforward to apply. What is often difficult is determining whether the function $f(z)$ is admissible for the method or not. Before we explain that notion, let's apply the theorem to $f(z) = e^z$, taking on faith, for now, the fact that it is admissible.

Example 1. (continued) The function e^z

First we would calculate $a(r) = r$, in this case, from (5.4.4). Then the equation $a(r_n) = n$, which determines $\{r_n\}$, becomes just $r_n = n$. These numbers r_n are the same as those that we found earlier from the condition (5.4.1). They are simply the values of r at which the minimum of $f(r)/r^n$ occurs.

Next we find $b(r) = r$ from (5.4.5), and Hayman's result (5.4.6) reads as

$$\frac{1}{n!} \sim \frac{e^n}{n^n \sqrt{2n\pi}},$$

which is Stirling's formula again, but this time in its exact form. ∎

Next let's give a precise definition of the class of admissible functions. Let $f(z) = \sum_{n\geq 0} a_n z^n$ be regular in $|z| < R$, where $0 < R \leq \infty$. Suppose that

(a) there exists an $R_0 < R$ such that

$$f(r) > 0 \qquad (R_0 < r < R),$$

and

(b) there exists a function $\delta(r)$ defined for $R_0 < r < R$ such that $0 < \delta(r) < \pi$ for those r, and such that as $r \to R$ uniformly for $|\theta| \leq \delta(r)$, we have

$$f(re^{i\theta}) \sim f(r)e^{i\{\theta a(r) - \frac{1}{2}\theta^2 b(r)\}},$$

and

(c) uniformly for $\delta(r) \leq |\theta| \leq \pi$ we have

$$f(re^{i\theta}) = \frac{o(f(r))}{\sqrt{b(r)}} \qquad (r \to R),$$

and

(d) as $r \to R$ we have $b(r) \to +\infty$, where $a(r)$, $b(r)$ are defined by (5.4.4), (5.4.5). Then we will say that $f(z)$ is *admissible*, and the apparatus of theorem 5.4.1 above is available for determining the asymptotic growth of the coefficients $\{a_n\}$.

However, it isn't always necessary to appeal directly to the definition of admissibility in order to be sure that a certain function is admissible. Here are some theorems that give sufficient conditions for admissibility, conditions that are much easier to verify than the formal definition above.

(A) If $f(z)$ is admissible, then so is $e^{f(z)}$.

(B) If f and g are admissible in $|z| < R$, then so is fg.

(C) Let f be admissible in $|z| < R$. Let P be a polynomial with real coefficients which satisfies $P(R) > 0$, if $R < \infty$, and which has a positive highest coefficient, if $R = +\infty$. Then $f(z)P(z)$ is admissible in $|z| < R$.

(D) Let P be a polynomial with real coefficients, and let f be admissible in $|z| < R$. Then $f + P$ is admissible, and if the highest coefficient of P is positive, then $P[f(z)]$ is also admissible.

(E) Let $P(z)$ be a nonconstant polynomial with real coefficients, and let $f(z) = e^{P(z)}$. If $[z^n]f(z) > 0$ for all sufficiently large n, then $f(z)$ is admissible in the plane.

Example 2.

Let t_n be the number of involutions of n letters, i.e., the number of permutations of n letters whose cycles have lengths ≤ 2. We will find the asymptotic behavior of $\{t_n\}$.

By (3.8.3), the egf of the sequence $\{t_n\}$ is

$$f(z) = \sum_{n \geq 0} \frac{t_n}{n!} z^n = e^{z + \frac{1}{2}z^2}.$$

By criterion (E) above, $f(z)$ is clearly Hayman admissible in the whole plane. Hence theorem 5.4.1 applies. To use it, we first calculate the functions $a(r)$, $b(r)$ of (5.4.4) and (5.4.5). We find that

$$a(r) = r\frac{f'(r)}{f(r)} = r + r^2$$

and

$$b(r) = ra'(r) = r + 2r^2.$$

Next we let r_n be the positive real solution of the equation $a(r_n) = n$, which in this case is the equation

$$r_n + r_n^2 = n. \qquad (5.4.7)$$

Evidently

$$
\begin{aligned}
r_n &= \sqrt{n + \frac{1}{4}} - \frac{1}{2} \\
&= \sqrt{n}\left\{1 + \frac{1}{4n}\right\}^{\frac{1}{2}} - \frac{1}{2} \\
&= \sqrt{n}\left\{1 + \frac{1}{8n} - \frac{1}{128n^2} + \cdots\right\} - \frac{1}{2} \qquad \text{(by (2.5.6))} \\
&= \sqrt{n} - \frac{1}{2} + \frac{1}{8\sqrt{n}} - \frac{1}{128n^{3/2}} + \cdots,
\end{aligned}
\qquad (5.4.8)
$$

so we have a very good fix on where r_n is, in this case.

Now, it would seem, all we have to do is to plug things into Hayman's estimate (5.4.6), and that's true, but there will be one little subtlety that will require a bit of explanation. If we take a look at (5.4.6) we see that we will need asymptotic estimates of the '$\sim$' kind for $f(r_n)$, $b(r_n)$, and r_n^n. Let's take them one at a time.

First,

$$f(r_n) = e^{r_n + \frac{1}{2}r_n^2} = e^{\frac{1}{2}(r_n + n)} = e^{n/2}e^{r_n/2},$$

where (5.4.7) was used again. But in view of (5.4.8),

$$
\begin{aligned}
e^{r_n/2} &= \exp\left\{\frac{\sqrt{n}}{2} - \frac{1}{4} + O(n^{-1/2})\right\}, \\
&\sim e^{\frac{1}{2}\sqrt{n} - \frac{1}{4}} \qquad (n \to \infty)
\end{aligned}
$$

and so

$$f(r_n) \sim \exp\left\{\frac{n}{2} + \frac{1}{2}\sqrt{n} - \frac{1}{4}\right\} \qquad (n \to \infty). \qquad (5.4.9)$$

So far, so good. Next on the list is $b(r_n)$, and that one is easy since

$$b(r_n) = r_n + 2r_n^2 \sim 2r_n^2 \sim 2n \qquad (n \to \infty). \qquad (5.4.10)$$

The last one is the hardest, and it is

$$
\begin{aligned}
r_n^n &= \left\{\sqrt{n} - \frac{1}{2} + \frac{1}{8\sqrt{n}} - \cdots\right\}^n \\
&= n^{\frac{n}{2}}\left\{1 - \frac{1}{2\sqrt{n}} + \frac{1}{8n} - \cdots\right\}^n.
\end{aligned}
\qquad (5.4.11)
$$

What we want to do now is to find just one term of the asymptotic behavior of the large curly brace to the nth power, and of course, it's that nth power that causes the difficulty.

To illustrate the method in a simpler context, consider $(1+\frac{1}{n})^n$. What does this behave like for large n? Does it approach 1? We know that it doesn't; in fact it approaches e. So the correct asymptotic relation is

$$\left(1+\frac{1}{n}\right)^n \sim e \qquad (n \to \infty).$$

Hence, although $1 + \frac{1}{n} \sim 1$, $(1+\frac{1}{n})^n \sim e$. In general, *one cannot raise both sides of an asymptotic equality to the nth power and expect it still to be true.* In exercise 7 below there are a number of situations of this kind to think about.

To take a slightly harder example, how would we deal with

$$\left(1+\frac{1}{\sqrt{n}}\right)^n ? \tag{5.4.12}$$

What does it behave like when n is large? The way to deal with all of these questions is first to replace $(1 + \cdots)^n$ by $\exp\{n \log(1 + \cdots)\}$. Next, the logarithm should be expanded by using the power series (2.5.2), to get

$$(1 + \cdots)^n = \exp\{n \log(1 + \cdots)\}$$
$$= \exp\left(n\{(\cdots) - (\cdots)^2/2 + (\cdots)^3/3 - \cdots\}\right).$$

The infinite series in the argument of the exponential must now be broken off at exactly the right place. *Terms can be ignored beginning with the first one which, when multiplied by n, still approaches 0.* More briefly, we can ignore all terms of that infinite series which are $o(n^{-1})$.

In the example (5.4.12) we have

$$\left(1+\frac{1}{\sqrt{n}}\right)^n = \exp\left\{n\log\left(1+\frac{1}{\sqrt{n}}\right)\right\}$$
$$= \exp\left\{n\left(\frac{1}{\sqrt{n}} - \frac{1}{2n} + O(n^{-3/2})\right)\right\}$$
$$\sim \exp\left\{n\left(\frac{1}{\sqrt{n}} - \frac{1}{2n}\right)\right\}$$
$$= \exp\left\{\sqrt{n} - \frac{1}{2}\right\}.$$

Now that we have that subject under our belts, we can return to the real problem, which is (5.4.11). We now find that

$$\left\{1 - \frac{1}{2\sqrt{n}} + \frac{1}{8n} - \cdots\right\}^n$$

$$= \exp\left\{n\log\left(1 - \frac{1}{2\sqrt{n}} + \frac{1}{8n} - \cdots\right)\right\}$$

$$= \exp\left\{n\left(\left(-\frac{1}{2\sqrt{n}} + \frac{1}{8n}\right) - \frac{1}{2}\left(-\frac{1}{2\sqrt{n}} + \frac{1}{8n}\right)^2 + O(n^{-3/2})\right)\right\}$$

$$\sim \exp -(\sqrt{n}/2).$$

Hence, from (5.4.11),

$$r_n^n \sim n^{n/2} \exp -(\sqrt{n}/2). \tag{5.4.13}$$

That finishes the estimation of the three quantities that are needed by Hayman's theorem. The result, obtained by putting (5.4.9), (5.4.10), and (5.4.13) into (5.4.6) is that

$$a_n = \frac{t_n}{n!} \sim \frac{e^{\frac{n}{2} + \sqrt{n} - \frac{1}{4}}}{2n^{\frac{n}{2}}\sqrt{n\pi}}.$$

Finally, if we multiply by $n!$ and use Stirling's formula, we obtain, for the number of involutions of n letters,

$$t_n \sim \frac{1}{\sqrt{2}} n^{n/2} \exp\left(-\frac{n}{2} + \sqrt{n} - \frac{1}{4}\right). \tag{5.4.14}$$

Exercises

1. Use the LIF to show that the (infinite) binomial coefficient sum

$$\xi = \sum_s \binom{sL+1}{s} \frac{A^{-sL-1}}{(sL+1)},$$

for $A > 1$ and integer $L > 0$, satisfies $\xi^L - A\xi + 1 = 0$.

2. The Legendre polynomials $\{P_n(x)\}$ are generated by

$$\frac{1}{\sqrt{1 - 2xt + t^2}} = \sum_{n \geq 0} P_n(x)t^n.$$

Let x be a fixed complex number that lies outside the real interval $[-1, 1]$, and let τ denote that one of the two roots of the equation $\tau^2 - 2x\tau + 1 = 0$ which is > 1 in absolute value. Use the method of Darboux to show that, as $n \to \infty$,

$$P_n(x) \sim \frac{\tau^{n+1}}{\sqrt{n\pi(\tau^2 - 1)}}.$$

3. If $u = u(t)$ satisfies $u = t\phi(u)$ and $n \geq 0$, show that

$$[u^n]\{\phi(u)\}^n = [t^n]\left\{ \frac{tu'(t)}{u(t)} \right\} = [t^n]\frac{1}{(1 - t\phi'(u(t)))}.$$

4. Define, for all $n \geq 0$, $\gamma_n = [x^n](1 + x + x^2)^n$.

   (a) Use the result of exercise 3 above to prove that for $n \geq 0$,

$$\gamma_n = [x^n]\left\{ \frac{1}{\sqrt{1 - 2x - 3x^2}} \right\}.$$

   (b) Show that, using the notation of problem 2 above,

$$\gamma_n = \left(\sqrt{3}/i \right)^n P_n(i/\sqrt{3}),$$

   and so obtain the asymptotic behavior of the sequence $\{\gamma_n\}$ for large n.

5. Define, for integer $p \geq 3$,

$$S_p(n) = \sum_{k=0}^{n} \binom{pn}{k} \qquad (n \geq 0).$$

(a) Exhibit $S_p(n)$ as $[x^n]$ in a certain ordinary power series, which (alas!) itself depends on n.

(b) Nevertheless, use the LIF (backwards) to show that

$$\sum_n S_p(n)x^n(1+x)^{-pn-1} = \frac{1}{(1-x)(1-(p-1)x)}.$$

(c) Deduce from part (b) that the $\{S_p(n)\}$ satisfy the recurrence

$$\sum_k (-1)^k \binom{pn-(p-1)k}{k} S_p(n-k) = \frac{(p-1)^{n+1}-1}{p-2} \qquad (n \geq 0).$$

(d) If $F(u) = \sum_{n\geq 0} S_p(n)u^n$, let

$$x = \frac{1}{(p-1)} - \epsilon$$

in part (b) to show that

$$F\left(\frac{(p-1)^{p-1}}{p^p} \left\{ 1 - \frac{(p-1)^3}{2p}\epsilon^2 + \cdots \right\} \right) = \frac{p}{(p-1)(p-2)\epsilon} + O(1)$$

as $\epsilon \to 0$.

(e) If

$$g(x) = F\left(\frac{(p-1)^{p-1}}{p^p} x \right)$$

then show that

$$g(x) = \frac{1}{(p-2)} \sqrt{\binom{p}{2}} \frac{1}{\sqrt{1-x}} + O(1).$$

(f) Use Darboux's method to show that, as $n \to \infty$,

$$S_p(n) \sim \frac{1}{(p-2)} \sqrt{\frac{\binom{p}{2}}{n\pi}} \left(\frac{p^p}{(p-1)^{p-1}} \right)^n.$$

(g) From part (b) show that

$$\sum_{n\geq 0} S_3(n) \left(\frac{4u^2}{27} \right)^n = \frac{u}{u - 2\sin(\frac{1}{3}\sin^{-1}u)} - \frac{2u}{2u - 3\sin(\frac{1}{3}\sin^{-1}u)}.$$

6. Under what additional conditions on a polynomial P with nonnegative
real coefficients will there exist an N such that for all $n > N$ we have
$[z^n]e^{P(z)} > 0$?

7. Find the asymptotic behavior (main term) of $(1 + \epsilon_n)^n$ if

  (a) $\epsilon_n = n^a \quad (0 < a < 1)$,

  (b) $\epsilon_n = n^{-a} \quad (0 < a < 1)$,

  (c) $\epsilon_n = n^{-a}\log n \quad (1 < a < 2)$.

8. The purpose of this problem is to find the asymptotic behavior of the
number a_n of permutations of n letters whose cycles are all of lengths ≤ 3,
by using Hayman's method and the Lagrange Inversion Formula. (The use
of a symbolic manipulation package on a computer is recommended for this
problem, in order to help out with some fairly tedious calculations with
power series that will be necessary).) The egf of $\{a_n\}$ is

$$f(z) = \exp\left\{z + \frac{z^2}{2} + \frac{z^3}{3}\right\}.$$

  (a) Show that f is admissible in the plane.

  (b) Because r_n in this case satisfies a *cubic* equation rather than a
      quadratic, as in the example in the text, we will use the LIF to
      find the root and its powers with sufficient precision. Show that
      if we write

$$u = 1/r_n; \quad t = n^{-1/3}; \quad \phi(u) = (1 + u + u^2)^{1/3},$$

      then u satisfies the equation $u = t\phi(u)$, which is in the form
      (5.1.1).

  (c) Use the LIF to show that the root r_n has the asymptotic expansion

$$\frac{1}{r_n} = \frac{1}{n^{1/3}} + \frac{1}{3}\frac{1}{n^{2/3}} + \frac{1}{3}\frac{1}{n} + \frac{8}{81}\frac{1}{n^{4/3}} + O(n^{-5/3}).$$

  (d) Explain why the number of terms that were retained in part (c) is
      the minimum number that can be retained and still get the first
      term of the asymptotic expansion of a_n with this method.

  (e) Show that

$$\frac{1}{r_n^n} \sim n^{-\frac{n}{3}}\exp\left\{\frac{1}{3}n^{2/3} + \frac{5}{18}n^{1/3}\right\}.$$

  (f) Show that

$$b(r_n) \sim 3n.$$

(g) Show that

$$f(r_n) \sim \exp\left\{\frac{1}{3}n + \frac{1}{6}n^{2/3} + \frac{5}{9}n^{1/3} - \frac{29}{162}\right\}.$$

(h) Combine the results of (d), (e), (f) to show that the number of permutations of n letters that have no cycles of lengths > 3 is

$$a_n \sim \frac{n^{\frac{2n}{3}}}{\sqrt{3}} \exp\left\{-\frac{2n}{3} + \frac{1}{2}n^{2/3} + \frac{5}{6}n^{1/3} - \frac{29}{162}\right\}.$$

9. Derive the power series expansion (2.5.16).

10. In this exercise, $\sigma(n,k)$ is the number of involutions of n letters that have exactly k cycles, and $t_n = \sum_k \sigma(n,k)$ is the number of involutions of n letters.

(a) Show that

$$\sum_{n,k} \frac{\sigma(n,k)}{n!} x^n y^k = e^{y(x + \frac{1}{2}x^2)}.$$

(b) Hence find the formula

$$\sigma(n,k) = \frac{n!}{(n-k)!(2k-n)!2^{n-k}}$$

for $\sigma(n,k)$.

(c) Using the results of part (a) and problem 5 of chapter 3, show that the *average* number of cycles in an involution of n letters is exactly

$$\frac{n}{2}\left\{1 + \frac{t_{n-1}}{t_n}\right\}.$$

(d) Using (5.4.14), show that the average number of cycles in an involution of n letters is

$$= \frac{n}{2} + \frac{1}{2}\sqrt{n}\,(1 + o(1)) \qquad (n \to \infty).$$

Appendix
Using Maple* and Mathematica**

Many branches of mathematics that were formerly thought of as being fit only for humans, are being invaded by computers. First, elementary school students learned how to multiply numbers with many digits and then found out that little calculators could do it for them. Other kinds of mathematics that are taught in secondary schools that now can be done by computers include expanding and factoring algebraic expressions, solving linear and quadratic equations, plotting graphs of curves and surfaces, doing logarithms and powers, and more.

At the university level we find now that "computer algebra" programs can differentiate functions symbolically, do integrals, vector analysis, linear algebra, etc., all *symbolically*, rather than numerically.

Here we want to show how computers can easily handle much of the routine work that is involved in solving problems about generating functions. To emphasize this point, we will show how well computers can do some of the homework problems in this book! Very well indeed, we're sure you will agree.

In this brief Appendix we'll discuss first how computer programs can do extensive manipulations of power series. Next we'll focus on one such program, *Mathematica*™ (Version 2.0) , and tell you about its amazing built-in `RSolve` function. Finally we will look at how *Maple*™ handles asymptotics, which can be quite a boon for problems such as those we looked at in the previous chapter.

1. Series manipulation

In *Mathematica*™, the instruction `Series[f,x,x0,m]` will display the first $m + 1$ terms of the power series expansion of f about $x = x0$. Thus, to see the first 10 terms of the series for $\sin x/(1 + x)$, about the origin, you would enter (the *Maple*™ instruction that would accomplish the same thing would be `series(sin(x)/(1+x),x=0,9)`)

$$\texttt{Series[Sin[x]/(1+x),\{x,0,9\}]}$$

and *Mathematica*™ would respond

$$x - \frac{7\,x^3}{6} + \frac{47\,x^5}{40} - \frac{5923\,x^7}{5040} + \frac{426457\,x^9}{362880} + \mathrm{O}(x)^{10}.$$

Perhaps you'd like to check the accuracy of the terms displayed in the series (2.5.10) of Chapter 2, and to see what the next two terms are. If so, then enter

---

*   *Maple* is a registered trademark of Waterloo Maple Software.
**   *Mathematica* is a registered trademark of Wolfram Research, Inc.

$$\texttt{Series[(1-Sqrt[1-4x])/(2x),}\{\texttt{x,0,9}\}\texttt{]}$$

and you will see

$$1 + x + 2\,x^2 + 5\,x^3 + 14\,x^4 + 42\,x^5 + 132\,x^6 + 429\,x^7 + 1430\,x^8$$
$$+ 4862\,x^9 + 16796\,x^{10} + 58786\,x^{11} + \mathrm{O}(x)^{12}.$$

If you want to obtain the list of coefficients of the terms of this series, because they are the numbers that the series "generates," then ask for

$$\texttt{CoefficientList[\%,x]}$$

to obtain (the "%" means the result of the computation in the preceding line)

$$\{1, 1, 2, 5, 14, 42, 132, 429, 1430, 4862, 16796, 58786\}$$

and there are the Catalan numbers on display.

If you want to see only the coefficient of x^7 then you would enter

$$\texttt{Coefficient[\%,x,7]}$$

instead, and the 429 would appear.

A little more work is needed to see sequences that are generated by exponential generating functions. Suppose you wanted the first 12 Bell numbers. According to theorem 1.6.1 these are the coefficients of $x^n/n!$ in

$$\texttt{Series[Exp[Exp[x]-1],}\{\texttt{x,0,12}\}\texttt{]}.$$

If you type exactly that, $Mathematica^{\mathrm{TM}}$ will reply with

$$1 + x + x^2 + \frac{5\,x^3}{6} + \frac{5\,x^4}{8} + \frac{13\,x^5}{30} + \frac{203\,x^6}{720} + \frac{877\,x^7}{5040} + \frac{23\,x^8}{224}$$
$$+ \frac{1007\,x^9}{17280} + \frac{4639\,x^{10}}{145152} + \frac{22619\,x^{11}}{1330560} + \frac{4213597\,x^{12}}{479001600} + \mathrm{O}(x)^{13},$$

which isn't quite what you wanted because, for instance, the coefficient of $x^8/8!$ is not readily apparent. One more instruction, such as

$$\texttt{Table[j! Coefficient[\%,x,j],}\{\texttt{j,0,12}\}\texttt{]}$$

will get the desired display of Bell numbers,

$$\{1,\ 1,\ 2,\ 5,\ 15,\ 52,\ 203,\ 877,\ 4140,\ 21147,\ 115975,\ 678570,\ 4213597\}.$$

2. The RSolve.m routine

The RSolve package was written in $Mathematica^{\mathrm{TM}}$ by Marko Petkovšek [Pe]. Its purpose is to find symbolic solutions to recurrence relations and difference equations. It can do so by explicitly finding the ordinary power series or exponential generating function of the unknown sequence.

To use it one first reads in the package with

$$\texttt{<<DiscreteMath/RSolve.m}$$

One then has a powerful facility for finding generating function solutions to problems in combinatorial recurrence.

Let's try it on the Fibonacci recurrence, with the call

```
RSolve[{f[n+2]==f[n+1]+f[n],f[0]==0,f[1]==1},f[n],n].
```

It replies, after an order to Simplify[%], as follows.

$$\{\{f(n) \to \frac{\left(-\left(\frac{1}{2} - \frac{\sqrt{5}}{2}\right)^n + \left(\frac{1}{2} + \frac{\sqrt{5}}{2}\right)^n\right) \, \text{If}(n \geq 1, 1, 0)}{\sqrt{5}}\}\},$$

which is, of course, the explicit formula for the Fibonacci numbers. If you're ready for this, let's change the call above by replacing "RSolve" by "GeneratingFunction," and adding one more argument, x say, to tell it the variable to use in the generating function. That means that we enter the request

```
GeneratingFunction[{f[n+2]==f[n+1]+f[n],f[0]==0,f[1]==1},f[n],n,x].
```

And what is the reply? It is

$$\{\{\frac{x}{1 - x - x^2}\}\},$$

which even in an age of multitudinous computer miracles must leave us in awe.

Perhaps you'd rather have the exponential generating function of your numbers. Well then you would change the call to

```
ExponentialGeneratingFunction[{f[n+2]==f[n+1]+f[n],f[0]==0,f[1]==1},f[n],n,x]
```

and the computer would inform you that

$$\{\{\frac{-e^{\frac{(1-\sqrt{5})\,x}{2}} + e^{\frac{(1+\sqrt{5})\,x}{2}}}{\sqrt{5}}\}\}$$

is the function you seek.

Now let's watch it solve the recurrence (2.2.6) for the number of block fountains of coins that have k coins in the first row. This time the call is

```
GeneratingFunction[f[k]==1+Sum[(k-j) f[j],{j,1,k}]/;k>=1, f[k],k,t],
```

and the response is

$$\{\{-\frac{(-1+t)\,t}{1 - 3\,t + t^2}\}\}$$

in agreement with (2.2.7). It can even find a closed formula for the number of such fountains from the generating function. To get that, ask for

```
Simplify[SeriesTerm[%,{t,0,n}]]
```

and the output will be

$$\{\{\left(\frac{(5 - \sqrt{5})\left(\frac{3}{2} + \frac{\sqrt{5}}{2}\right)^n}{10} + \frac{\left(\frac{3}{2} - \frac{\sqrt{5}}{2}\right)^n (5 + \sqrt{5})}{10}\right) \, \text{If}(n \geq 0, 1, 0)$$

$$- \text{If}(n = 0, 1, 0)\}\}.$$

As you can see, it did the partial fraction expansion followed by two geometric series manipulations, just as we did to obtain, for instance, (1.3.3).

The package can also find closed form expressions for the sums of series in which formulas are given for the nth coefficient. A request

$$\texttt{PowerSum[a n+b,\{z,n\}]}$$

will produce the answer to exercise 1(b) in this book, in the form

$$\frac{b}{1-z} + \frac{a\,z}{(-1+z)^2}.$$

It can do much harder ones than that, like the gf of the harmonic numbers that we did in Example 5 of chapter 2. That one is the answer to the call

$$\texttt{PowerSum[Sum[1/j,\{j,1,n\}],\{x,n\}],}$$

namely

$$-\frac{\log(1-x)}{1-x}.$$

The reader who takes the time to experiment with the capabilities of the `RSolve.m` package will be amply rewarded.

3. Asymptotics in *Maple™*

In *Maple™*, if you type `asympt(f,x,n);` you will receive n terms of the asymptotic expansion of the function f of the variable x, as $x \to \infty$. Let's try Stirling's formula first, by asking for

$$\texttt{asympt(n!,n,5);}$$

The computer's answer is (we use 'Pi' instead of 'π' etc. because that's pretty much how it will look on your screen)

$$\left(2^{1/2}Pi^{1/2}n^{1/2} + 1/12\frac{2^{1/2}Pi^{1/2}}{n^{1/2}} + 1/288\frac{2^{1/2}Pi^{1/2}}{n^{3/2}} - \frac{139}{51840}\frac{2^{1/2}Pi^{1/2}}{n^{5/2}}\right.$$
$$\left. - \frac{571}{2488320}\frac{2^{1/2}Pi^{1/2}}{n^{7/2}} + O(\frac{1}{n^{9/2}})\right)/((1/n)^n exp(n)).$$

We all know that $(1+1/n)^n \to e$, but how fast does it go? The answer given by *Maple™* is

$$exp(1) - 1/2\frac{exp(1)}{n} + \frac{11}{24}\frac{exp(1)}{n^2} + O(\frac{1}{n^3}).$$

In closing, let's do exercise 8(d) of the previous chapter, which asks for the asymptotic behavior of the nth power of

$$\frac{1}{r_n} = \frac{1}{n^{1/3}} + \frac{1}{3}\frac{1}{n^{2/3}} + \frac{1}{3}\frac{1}{n} + \frac{8}{81}\frac{1}{n^{4/3}} + O(n^{-5/3}).$$

Needless to say, *Maple*™ is up to the task, and gives

$$n^{-\frac{n}{3}} \exp\left\{\frac{1}{3}n^{2/3} + \frac{5}{18}n^{1/3}\right\}(1 + O(1)).$$

Exercises

On any computer that is available to you, do the following.

1. Exercises 1, 2, 5, 6, 8 of Chapter 1.

2. Check the first five terms of any five of the series displayed in section 2.5.

3. Exercises 1, 2, 4 of chapter 2.

4. Use the "series" command to find the first 15 values of $g(n)$ of (3.9.1).

5. From (3.8.3), tabulate the number of involutions of n letters, for $n \leq 15$.

6. Use the asymptotics capability of *Maple*™ to find the first 5 terms of the asymptotic expansions of the following.

   (a) $(1 + 1/\sqrt{n})^n$

   (b) $\sqrt{n!}$

   (c) $(1 + 1/n)^{\sqrt{n}}$

   (d) $\sin(\sin 1/x)$

Solutions

Answers to problems for chapter 1

1.

    (a) $(xD)(1/(1-x)) = x/(1-x)^2$

    (b) $(\alpha xD + \beta)(1/(1-x)) = \alpha x/(1-x)^2 + \beta/(1-x)$

    (c) $(xD)^2(1/(1-x))$

    (d) $(\alpha(xD)^2 + \beta xD + \gamma)(1/(1-x))$

    (e) $P(xD)(1/(1-x))$

    (f) $1/(1-3x)$

    (g) $5/(1-7x) - 3/(1-4x)$

2.

    (a) $(xD)e^x = xe^x$

    (b) $(\alpha xD + \beta)e^x = (\alpha x + \beta)e^x$

    (c) $(xD)^2 e^x = (x + x^2)e^x$

    (d) $(\alpha(xD)^2 + \beta xD + \gamma)e^x$

    (e) $P(xD)e^x$

    (f) e^{3x}

    (g) $5e^{7x} - 3e^{4x}$

3.

    (a) $f(x) + c/(1-x)$

    (b) $\alpha f(x) + c/(1-x)$

    (c) $xDf(x)$

    (d) $P(xD)f(x)$

    (e) $f(x) - a_0$

    (f) $f(x) - a_0 - a_1 x + (1 - a_2)x^2$

    (g) $(f(x) + f(-x))/2$

    (h) $(f(x) - a_0)/x$

(i) $(f(x) - \sum_0^{h-1} a_j x^j)/x^h$

(j) $(f - a_0 - a_1 x)/x^2 + 3((f - a_0)/x) + f$

(k) $(f - a_0 - a_1 x)/x^2 - ((f - a_0)/x) - f$

4.

(a) $f(x) + ce^x$ (b) $\alpha f(x) + ce^x$ (c) $xf'(x)$

(d) $P(xD)f(x)$ (e) $f - a_0$ (f) $f - a_0 - a_1 x + (1 - a_2)x^2/2$

(g) $(f(x) + f(-x))/2$ (h) $f'(x)$ (i) $D^h f(x)$

(j) $f'' + 3f' + f$ (k) $f'' - f' - f$

5.

(a) $2^n/n!$

(b) α^n

(c) $(-1)^m$ if $n = 2m + 1$ is odd, and 0 else.

(d) $(a^{n+1} - b^{n+1})/(a - b)$

(e) $\binom{m}{n/2}$

6.

(a) We see at once that $f/x = 3f + 2/(1-x)$, so $f = 2x/((1-x)(1-3x))$.

(b) $f/x = \alpha f + \beta/(1 - x)$ so $f = \beta x/((1 - x)(1 - \alpha x))$.

(c) Here $(f - x)/x^2 = 2f/x - f$ so $f = x/(1 - x)^2$.

(d) Since $f/x = f/3 + 1/(1 - x)$ we have $f = 3x/((1 - x)(3 - x))$.

8.

(a) $f' = 3f + 2e^x$, $f(0) = 0$ give $f = e^{3x} - e^x$

(b) $f' = \alpha f + \beta e^x$ so $f = (\beta/(1 - \alpha))(e^x - e^{\alpha x})$

(c) $f'' = 2f' - f$, $f(0) = 0$, $f'(0) = 1$ yield $f = xe^x$

(d) $f' = f/3 + e^x$, $f(0) = 0$ give $f = \frac{3}{2}(e^x - e^{x/3})$

9. Multiply both sides of the equation $f(2n) = f(n)$ by x^{2n} and sum over $n \geq 1$. Then multiply both sides of $f(2n + 1) = f(n) + f(n + 1)$ by x^{2n+1}, sum over $n \geq 1$, and add to the previous result. Then add $f(1)x = x$ to that result to obtain the functional equation.

To find the explicit infinite product form of the solution, let's first see how we might guess that answer, and then how we might prove it.

Take the functional equation for F, and replace x by x^2 throughout, then substitute the result back in the functional equation, to get

$$F(x) = (1 + x + x^2)(1 + x^2 + x^4)F(x^4).$$

If we now replace x by x^2 again, and substitute we'll get even more factors of the infinite product. Hence we should suspect that the product is the answer. To *prove* that the product is the answer, we have two choices. First, over the ring of formal power series, consider the product as a formal beast which obviously satisfies the functional equation for F. Second, analytically, an infinite product $\prod(1+q_n)$ converges if the series $\sum |q_n|$ does; so, the product converges for $|x|$ small enough, to an analytic function F.

10. For part (a) see section 4.1.

(b) $p_n^{(2)} = \sum_{j=0}^{n} \text{Prob}(X = j)\text{Prob}(X = n - j) = [x^n]P(x)^2$.

(c) $P_k(x) = P(x)^k$

(d) By part (c) the mean is

$$P_k'(1)/P_k(1) = \left[kP(x)^{k-1}P'(x)/P(x)^k\right]_{x=1} = k\mu,$$

and the variance is

$$(\log P_k(x))' + (\log P_k(x))''\bigg|_{x=1} = k(\log P(x))' + (\log P(x))''\bigg|_{x=1},$$

which is $k\sigma^2$.

(e) Since $A^k = B$ we have $kA'/A = B'/B$ or $kA'B = AB'$. Equate the coefficients of x^n to find that $nb_n = \sum_{j=1}^{n}(j(k+1) - n)a_j b_{n-j}$ for $n \geq 1$, with $b_0 = 1$.

(f) p^* is the coefficient of x^{300} in $(.1x + .2x^2 + .1x^3 + .2x^4 + .2x^5 + .2x^6)^{100}/(1 - x)$. Numerically, it is about $.00000095$.

(g) The required probability is

$$[x^j]\frac{1}{1-x}\left(\frac{x + x^2 + \cdots + x^m}{m}\right)^n = \frac{1}{m^n}[x^{j-n}]\frac{(1 - x^m)^n}{(1 - x)^{n+1}}.$$

The result follows by expanding the numerator by the binomial theorem, the reciprocal of the denominator by the binomial series, and multiplying.

11. Among these subsets we distinguish those that do contain n and those that don't. If such a subset contains n, then the rest of that subset is one of the subsets that is counted by $f(n - 2)$. If it does not contain n then the entire subset is one of those that is counted by $f(n - 1)$. Thus $f(n) = f(n-1)+f(n-2)$, which together with the starting values $f(1) = 2$, $f(2) = 3$ tells us that $f(n) = F_{n+2}$, where the F's are the Fibonacci numbers.

12. As in problem 11, we distinguish those k-subsets that do contain n and those that do not. If such a k-subset does contain n then the rest of that subset is one of the subsets that is counted by $f(n - 2, k - 1)$, otherwise the entire k-subset is one of those that is counted by $f(n-1,k)$. Thus $f(n,k) = f(n-1,k) + f(n-2,k-1)$, for $k \geq 2$. If we define $F_k(x) = \sum_{n \geq 1} f(n,k)x^n$, then after multiplying the recurrence by x^n and summing over $n \geq 1$ we find that $F_k(x) = x^2 F_{k-1}(x)/(1-x)$, which together with $F_1(x) = x/(1-x)^2$ tells us that $F_k(x) = x^{2k-1}/(1-x)^{k+1}$. If we expand in the binomial series we find that

$$f(n,k) = [x^n]F_k(x) = [x^n]x^{2k-1}\sum_{h \geq 0}\binom{k+h}{k}x^h = \binom{n-k+1}{k}.$$

13. From problems 11 and 12, it must be that

$$\sum_k \binom{n-k+1}{k} = F_{n+1} \qquad (n \geq 0),$$

which will be proved another way in example 1 of section 4.3.

14. A circular arrangement that does contain n is obtained by taking a linear arrangement of $\{2,3,\ldots,n-2\}$, no two consecutive (on the line), adjoining n to it, and laying it out around a necklace. So by exercise 11, there are F_{n-1} such arrangements. One that does *not* contain n is obtained by taking any linear arrangement of $\{1,2,\ldots,n-1\}$ and laying it out around a necklace, so there are F_{n+1} of these. Hence there are $F_{n-1} + F_{n+1}$ such circular sequences altogether.

15. As in the previous problem, the answer is $f(n-3,k-1) + f(n-1,k)$ where $f(n,k) = \binom{n-k+1}{k}$ is the solution to exercise 12.

16. The partial fraction expansion of $1/(1-x^2)^2$ is

$$\frac{1}{4(1-x)^2} + \frac{1}{4(1-x)} + \frac{1}{4(1+x)^2} + \frac{1}{4(1+x)}.$$

Therefore the coefficient of x^n in its power series expansion is

$$\frac{n+1}{4} + \frac{1}{4} + \frac{(-1)^n(n+1)}{4} + \frac{(-1)^n}{4},$$

which is 0 if n is odd and is $(n+2)/2$ if n is even. Otherwise, take the series for $1/(1-u)^2$ and replace u by x^2. Even sneakier would be to use one of the symbolic manipulation programs that are now available on computers. They can produce the general term of such series on demand.

17. Fix j, $1 \leq j \leq n$, and consider just those permutations σ of n letters that have $\sigma(j) = n$. Then no inversions have j as the second member of the pair and exactly $n-j$ inversions have j as the first member of the pair. Hence if we delete n from the string of values of σ we obtain a permutation of $n-1$ letters with $n-j$ fewer inversions. Thus $b(n,k) = \sum_j b(n-1,k-n+j)$. Multiply by x^k and sum on k to obtain $B_n(x) = (1+x+\cdots+x^{n-1})B_{n-1}(x)$. Hence $b(n,k)$ is the coefficient of x^k in

$$(1+x)(1+x+x^2)(1+x+x^2+x^3)\cdots(1+x+x^2+x^3+\cdots+x^{n-1}).$$

18.

(a) The probability is evidently $1/k!$ that the first k values will decrease, so $n!/k!$ of the permutations have this property.

(b) The probability that a permutation begins with k decreasing values followed by an increasing one is $1/k!-1/(k+1)!$, if $0 \le k < n$, and is $1/n!$ when $k=n$. The average value of k, weighted with these probabilities, is

$$\sum_{k=0}^{n-1} k\left(\frac{1}{k!} - \frac{1}{(k+1)!}\right) + \frac{n}{n!} = \sum_{k=1}^{n} \frac{1}{k!},$$

and therefore an average permutation of n letters begins with a decreasing sequence whose length is approximately $e-1$.

(c) If we begin with a permutation of $n-1$ letters that has k runs, then by inserting the letter n in each of the n possible places we manufacture k permutations of n letters that have k runs and $n-k$ permutations of n letters that have $k+1$ runs. Thus $f(n,k) = kf(n-1,k) + (n-k+1)f(n-1,k-1)$.

19.

(a) It is $(1+x)^2(1+x^2)(1+x^5)(1+x^{10})^2(1+x^{20})(1+x^{50})$.

(b) The sum represents $3^8 = 6561$ integers, each between -99 and 99. Hence these 199 integers are represented an average of $6561/199 = 32.9..$ ways, so some integer must be represented at least 33 ways. The required product is

$$(1/x+1+x)^2(1/x^2+1+x^2)(1/x^5+1+x^5)\cdots(1/x^{50}+1+x^{50}).$$

(c) If $w_1, \cdots, w_r$ are distinct integers, and if D_n is the number of representations of n as a sum $n = w_1x_1 + w_2x_2 + \cdots + w_kx_k$ where each of the x_i is ± 1, then

$$\sum_{n} D_n t^n = \prod_{i=1}^{r}(t^{w_i} + t^{-w_i}).$$

The set of roots is the union of the sets of $2w_i$th roots of -1, for $i = 1, \ldots, k$.

Answers to problems for chapter 2

1. The thing to remember is that $1/(1-u) = 1 + u + u^2 + \cdots$.

(a) We have

$$\frac{1}{\cos x} = \frac{1}{1 - (\frac{x^2}{2} - \frac{x^4}{24} + \cdots)}$$

$$= 1 + (\frac{x^2}{2} - \frac{x^4}{24} + \cdots) + (\frac{x^2}{2} - \frac{x^4}{24} + \cdots)^2 + \cdots$$

$$= 1 + \frac{x^2}{2} + \frac{5x^4}{24} + \cdots$$

(b) Here we use the binomial theorem with negative exponent.

$$\frac{1}{(1+x)^m} = (1+x)^{-m}$$

$$= \sum_k \binom{-m}{k} x^k$$

$$= 1 + \binom{-m}{1} x + \binom{-m}{2} x^2 + \binom{-m}{3} x^3 + \cdots$$

$$= 1 - mx + \frac{m(m+1)}{2} x^2 - \frac{m(m+1)(m+2)}{6} x^3 + \cdots$$

(c) This is like part (a). We find

$$\frac{1}{1 + (t^2 + t^3 + t^5 + \cdots)} = 1 - (t^2 + t^3 + t^5 + \cdots) + \cdots$$

$$= 1 - t^2 - t^3 + t^4 + t^5 + \cdots.$$

2. In each case (except (e)), replace x by $x + bx^2 + cx^3 + \cdots$, set the result equal to x, and equate the coefficients of like powers of x to 0 to solve for b and c.

(a) $x + \frac{x^3}{6} + \cdots$

(b) $x - \frac{x^3}{3} + \cdots$

(c) $x - x^2 + \frac{3}{2}x^3 + \cdots$

(d) $x - x^3 + \cdots$

(e) In this part, note that if y is the inverse function, then

$$\log(1 - y) = x,$$

i.e., $y = 1 - e^x = -x - x^2/2 - x^3/6 - \cdots$.

3. If $f = \sum_{k \geq 0} a_k x^k$ then

$$0 = f'' + f = \sum_{k \geq 0}\{(k+2)(k+1)a_{k+2} + a_k\}x^k$$

and so

$$a_{k+2} = -\frac{a_k}{(k+1)(k+2)} \qquad (k = 0, 1, 2, \ldots).$$

If a_0 and a_1 are arbitrarily fixed, then by induction on k

$$a_{2k} = (-1)^k a_0/(2k)!$$

and

$$a_{2k+1} = (-1)^k a_1/(2k+1)!$$

for all $k \geq 0$, and the result follows.

4.

(a) $\frac{x}{(1-x)^2} + \frac{7}{1-x}$

(b) $\frac{x^4}{1-x}$

(c) $\frac{1}{1-x^2}$

(d) $\{\log\left(\frac{1}{1-x}\right) - x - \frac{x^2}{2}\}/x$

(e) $\{e^x - 1 - x - x^2/2 - x^3/6 - x^4/24\}/x^5$

(f) $x\frac{d}{dx}\{\frac{x}{1-x-x^2}\} = \frac{x(1+x^2)}{(1-x-x^2)^2}$

(g) $\{(xDxD) + (xD) + 1\}(e^x - 1) = (1+x)^2 e^x - 1$

5. In the binomial theorem $(1 + x)^n = \sum_k \binom{n}{k}x^k$, let $x = 1$.

6. We have

$$f(n, k) = \sum_{n_1 + \cdots + n_k = n} n_1 n_2 \cdots n_k$$

$$= [x^n](\sum_r r x^r)^k$$

$$= [x^n]\left\{\frac{x}{(1-x)^2}\right\}^k$$

$$= [x^n]\frac{x^k}{(1-x)^{2k}}.$$

Hence $\sum_n f(n,k)x^n = \frac{x^k}{(1-x)^{2k}}$. Explicitly, since

$$\frac{1}{(1-x)^{2k}} = \sum_{r\geq 0}\binom{r+2k-1}{r}x^r,$$

we find that $f(n,k) = \binom{n+k-1}{n-k}$. Notice that the answer is 0 when $n < k$. Explain why.

7. As in problem 6 above,

$$f(n,k,h) = [x^n]\left\{\sum_{r\geq h}x^r\right\}^k$$

$$= [x^n]\left\{\frac{x^{kh}}{(1-x)^k}\right\}.$$

8. 1, 1, 1, 1, and 1, respectively.

11. 1, 1, $5^{-\frac{1}{2}}$, 0, 1, respectively.

13. Let a and b be relatively prime. Then every divisor d of ab is uniquely of the form $d = d'd''$ where $d'\backslash a$, $d''\backslash b$. Hence

$$g(ab) = \sum_{d\backslash ab} f(d) = \sum_{\substack{d'\backslash a\\d''\backslash b}} f(d'd'')$$

$$= \sum_{\substack{d'\backslash a\\d''\backslash b}} f(d')f(d'') = \left(\sum_{d'\backslash a}f(d')\right)\left(\sum_{d''\backslash b}f(d'')\right) = g(a)g(b).$$

14. Consider the n fractions $1/n, 2/n, \ldots, n/n$. If we write them in lowest terms, then each of them will reduce to a fraction h/k, where $k\backslash n$ and h, k are relatively prime. Further, for a fixed divisor k of n, each of the $\phi(k)$ such fractions h/k occurs in exactly one way, i.e., by reducing exactly one fraction m/n.

15. For Euler's function, apply Möbius inversion to the result of problem 14 above. This gives

$$\phi(n) = \sum_{d\backslash n}\mu(n/d)d = n\sum_{d\backslash n}\frac{\mu(n/d)}{n/d} = n\sum_{d\backslash n}\frac{\mu(d)}{d}.$$

Since μ is multiplicative, $\mu(n)/n$ is a multiplicative function of n, and so, by problem 13, is the last member above.

For $\sigma(n)$, suppose a, b are relatively prime. Then every divisor of ab is uniquely of the form $d'd''$, where d' and d'' are divisors of a and of b, respectively, and the result follows. Finally, if a, b are relatively prime, $|\mu(ab)| = 1$ iff ab is squarefree iff a and b are squarefree.

16.

    (a) $\zeta(s-1)/\zeta(s)$

    (b) $\zeta(s)\zeta(s-1)$

    (c) $\zeta(s)/\zeta(2s)$

17.

    (a) $\zeta(s-1)$

    (b) $\zeta(s-a)$

    (c) $-\zeta'(s)$

    (d) $\zeta(s)\zeta(s-q)$

18.

    (a) $\zeta(s)\{\zeta(s-1)/\zeta(s)\} = \zeta(s-1)$

    (b) $\zeta(s)\{1/\zeta(s)\} = 1$

    (c) $\{1/\zeta(s)\}\{\zeta^2(s)\} = \zeta(s)$

19. We find that

$$F(x) = \frac{1}{10}\left(\frac{5-\sqrt{5}}{1-\alpha_+ x} + \frac{5+\sqrt{5}}{1-\alpha_- x}\right)$$

where $\alpha_\pm = (3 \pm \sqrt{5})/2$. Hence there are exactly

$$\frac{5-\sqrt{5}}{10}\alpha_+^k + \frac{5+\sqrt{5}}{10}\alpha_-^k$$

block fountains whose first row contains k coins.

21.

    (a) $\sum_n f(n,k,T)x^n = (\sum_{t\in T} x^t)^k$

    (b) $\sum_n g(n,k,T)x^n = [y^k]\prod_{t\in T}(1+yx^t)$

    (c) $\sum_n f(n,k,S,T)x^n = [\frac{y^k}{k!}]\prod_{t\in T}\{\sum_{s\in S}\frac{y^s x^{st}}{s!}\}$

22. Check that the required number is the coefficient of x^n in

$$\sum_{k\geq 1}(-1)^k\left(\frac{x}{1-x}\right)^k = -x,$$

hence $f(n) = 0$ for all n except that $f(1) = -1$.

23. On the one hand,

$$\frac{x(e^{mx}-1)}{e^x-1} = \left(\frac{x}{e^x-1}\right)(e^{mx}-1)$$

$$= \left(\sum_n \frac{B_n}{n!}x^n\right)\left(\sum_{j\geq 1}\frac{m^j x^j}{j!}\right)$$

$$= \sum_{n\geq 0}\frac{x^n}{n!}\left\{\sum_{j\geq 1}\binom{n}{j}B_{n-j}m^j\right\}.$$

On the other hand,

$$\frac{x(e^{mx}-1)}{e^x-1} = x\left(\frac{e^{mx}-1}{e^x-1}\right)$$

$$= x(1 + e^x + e^{2x} + \cdots + e^{(m-1)x})$$

$$= x\sum_{j=0}^{m-1}\sum_{r\geq 0}\frac{j^r x^r}{r!}$$

$$= \sum_{r\geq 0}\frac{x^{r+1}}{r!}S_r(m-1),$$

where $S_r(m)$ is the sum of the rth powers of the integers $1,\ldots,m$. If we compare the coefficients of x^n we find the explicit formula

$$S_n(m) = \frac{1}{n+1}\sum_{r\geq 1}\binom{n+1}{r}B_{n+1-r}(m+1)^r,$$

which holds for integers $m, n \geq 1$. The first few cases are, for $n = 1, 2, 3$,

$$1 + 2 + \cdots + m = \frac{m^2}{2} + \frac{m}{2}$$

$$1^2 + 2^2 + \cdots + m^2 = \frac{m^3}{3} + \frac{m^2}{2} + \frac{m}{6}$$

$$1^3 + 2^3 + \cdots + m^3 = \frac{m^4}{4} + \frac{m^3}{2} + \frac{m^2}{4}.$$

25.

   (a) If they differ in the jth bit, then their colors differ by j which is not 0. If they differ in the jth and kth bits, then their colors differ by $j + k$ or $j - k$ modulo $2n$, neither of which can be 0.

   (c) $f(z) = \prod_{k=1}^n (1 + z^k)$.

27.

    (a) $e^{-x}/(1-x)$.

    (b) If $D(x) = e^{-x}/(1-x)$ then $(1-x)D' = D - e^{-x}$. Match $[x^n]$ on both sides.

    (c) If $D_1(n)$ is the number with one fixed point then $D_1(n) = nD(n-1)$. Thus $D(n) - D_1(n) = D(n) - nD(n-1) = (-1)^n$ by part (b).

    (d) To construct a permutation of n letters that has k fixed points, we can choose the k fixed points in $\binom{n}{k}$ ways, and the rest of the permutation in $D(n-k)$ ways. Hence $D_k(n) = \binom{n}{k}D(n-k)$. Now multiply by $x^n y^k/n!$, sum, and use part (a).

28. We have

$$\sum_{d\backslash n} \mu(n/d)a_d(x^{n/d}) = \sum_{d\backslash n} \mu(n/d) \sum_{\delta\backslash d} b_{d/\delta}(x^{n\delta/d})$$

in which the coefficient of $b_r(x^{n/r})$ is $\sum_{\delta\backslash n/r} \mu(n/(r\delta))$, which vanishes unless $r = n$ and is 1 in that case.

30.

    (a) $\sum_{n\geq 1} \sqrt{n}/n^s = \sum_{n\geq 1} 1/n^{s-1/2} = \zeta(s-1/2)$.

    (b) The function is multiplicative and its value at $n = p^a$ is 0 if $a \geq 2$ and 1 otherwise. Hence by (2.6.6) the generating function is

$$\prod_p \{1 + p^{-s}\} = \prod_p \frac{1 - p^{-2s}}{1 - p^{-s}} = \frac{\zeta(s)}{\zeta(2s)}.$$

    (c) Here $\lambda(p^a) = (-1)^a$ for all $a \geq 0$, so by (2.6.6) its generating function $\Lambda(s)$ is

$$\prod_p \{1 - p^{-s} + p^{-2s} - \cdots\} = \prod_p \frac{1}{1 + p^{-s}} = \frac{\zeta(2s)}{\zeta(s)}.$$

    Finally, equate coefficients of n^{-s} on both sides of $\Lambda(s)\zeta(s) = \zeta(2s)$.

31.

    (a) If

$$\sum_{n\geq 1} b_n x^n = \sum_{n\geq 1} a_n \frac{x^n}{1 - x^n} = \sum_n a_n \sum_{m\geq 1} x^{mn} = \sum_r x^r \sum_{d\backslash r} a_d$$

    thus $b_n = \sum_{d\backslash n} a_d$, just as in the theory of Dirichlet series.

(b) Apply part (a) with $a_n = \mu(n)$.

(c) It is

$$\sum_{n\geq 1} \phi(n)\frac{x^n}{1-x^n} = \sum_{n\geq 1} nx^n = \frac{x}{(1-x)^2}$$

since $\sum_{d\backslash n} \phi(d) = n$.

32.

(a) $f/(1-x)$

(b) $f/(1-x)^r$

(c) By part (b), it is the sequence whose gf is $1/(1-x)^{r+1}$, which, by (2.5.7), is the sequence $\binom{n+r}{n}_{n\geq 0}$.

(d) It is the coefficient of x^n in $(\sum a_n x^n)/(1-x)^r$, viz.

$$\sum_{m=0}^{n} \binom{m+r-1}{m} a_{n-m} \qquad (n=0,1,2,\ldots).$$

(e) Then $f(x)/(1-x)^r = 1$, so $f(x) = (1-x)^r$ and $a_n = \binom{r}{n}(-1)^n$ for $n \geq 0$.

33.

(b) We have

$$\Phi_{p^a}(x) = \prod_{d\backslash p^a} (1-x^d)^{\mu(p^a/d)}$$

$$= \frac{1-x^{p^a}}{1-x^{p^{a-1}}} = 1 + x^{p^{a-1}} + x^{2p^{a-1}} + \cdots + x^{(p-1)p^{a-1}}.$$

(c) Since $\prod_{m\backslash n} \Phi_m(x) = 1 - x^n$ we have

$$\prod_{\substack{m\backslash n \\ m>1}} \Phi_m(1) = n. \qquad (*)$$

Now for $n = p^k$ use induction on k. If n is not a prime power let p^a be the highest power of p that divides n. Then in $(*)$ above, each divisor p^j $(1 \leq j \leq a)$ contributes a factor of p, so all such divisors contribute p^a. But no higher power of p divides n, so $\Phi_n(1)$ cannot be divisible by p. Since p was arbitrary, $\Phi_n(1)$ must be ± 1, and it is easy to rule out -1.

34.

    (a) This says that each integer r, $1 \le r \le n$ is uniquely of the form $r = md$ where $d \backslash n$ and $\gcd(m, n/d) = 1$. But this is clear since we take $d = \gcd(r, n)$ and $m = r/d$.

    (c) Let $x \to \omega$ in the result of part (b), and use L'Hospital's rule.

Answers to problems for chapter 3

1. A partition of n into odd parts looks like

$$n = r_1 \cdot 1 + r_3 \cdot 3 + r_5 \cdot 5 + \cdots .$$

Now substitute the binary expansion of each r_i, to get

$$n = (2^{a_1} + 2^{b_1} + \cdots) \cdot 1 + (2^{a_3} + 2^{b_3} + \cdots) \cdot 3 + (2^{a_5} + 2^{b_5} + \cdots) \cdot 5 + \cdots .$$

But now we have a partition of n into distinct parts, viz.,

$$n = 2^{a_1} + 2^{b_1} + \cdots + 2^{a_3} \cdot 3 + 2^{b_3} \cdot 3 + \cdots + 2^{a_5} \cdot 5 + 2^{b_5} \cdot 5 + \cdots .$$

(What partition corresponds to 39=3+3+7+7+19 ?) The map is uniquely invertible.

2. The deck has a card corresponding to each cyclic permutation of length $k, 2k, 3k, \ldots$. The number of these on mk letters is $(mk - 1)!$. The deck enumerator is

$$\mathcal{D}(x) = \sum_{m \ge 1} \frac{(mk - 1)!}{(mk)!} x^{mk} = \sum_{m \ge 1} \frac{x^{mk}}{mk} = \frac{1}{k} \log \frac{1}{1 - x^k} .$$

Hence the hand enumerator, without regard to number of cards in the hand, is

$$\mathcal{H}(x) = \exp \left\{ \frac{1}{k} \log \frac{1}{1 - x^k} \right\}$$

$$= \frac{1}{(1 - x^k)^{1/k}}$$

$$= \sum_{m \ge 0} \binom{-\frac{1}{k}}{m} (-1)^m x^{mk} .$$

The required number is the coefficient of $x^n / n!$ here, which is 0 if k does not divide n, and is

$$(-1)^r \binom{-\frac{1}{k}}{r} n! = \frac{n!}{r! k^r} (k + 1)(2k + 1) \cdots ((r - 1)k + 1)$$

if $n/k = r$ is an integer.

3. It is $\exp\left\{\sum_p \frac{x^p}{p!}\right\}$.

4.

    (a) The order is the least common multiple of the cycle lengths.

    (b) Clearly,

$$\tilde{g}(n,k) = \sum_{d\backslash k} g(n,d).$$

    Hence by Möbius inversion (2.6.12) we have

$$g(n,k) = \sum_{d\backslash k} \mu(k/d)\tilde{g}(n,d).$$

5.

    (a) One finds by logarithmic differentiation of the egf (3.8.3) that

$$T_n = T_{n-1} + (n-1)T_{n-2} \qquad (n \geq 2; T_0 = 1; T_1 = 1).$$

    (b) 1, 2, 4, 10, 26, 76

    (c) Consider separately those involutions of n letters for which n is a fixed point and those for which n is not fixed.

6. The deck enumerator is

$$\mathcal{D}(x) = \sum_{n\geq 4} \frac{x^n}{n} = \log\frac{1}{(1-x)} - x - x^2/2 - x^3/3,$$

and so the hand enumerator is

$$\mathcal{H}(x) = \frac{e^{-x-x^2/2-x^3/3}}{(1-x)}.$$

7. A card of weight n is a path or a cycle. If $n \geq 3$, there are $(n-1)!/2$ 'cycle cards' of weight n, and if $n \geq 2$ there are $n!/2$ 'path cards.' Hence

$$\mathcal{D}(x) = \left(\frac{1}{(1-x)} - \log(1-x) + 1 - 2x - x^2/2\right)/2,$$

and the hand enumerator $\mathcal{H}(x)$ is

$$\sinh\left(\frac{1}{2(1-x)} - \frac{1}{2}\log(1-x) - \frac{1}{2} - x - \frac{x^2}{4}\right)$$
$$= 1\frac{x^2}{2!} + 4\frac{x^3}{3!} + 15\frac{x^4}{4!} + 72\frac{x^5}{5!} + 435\frac{x^6}{6!} + 3300\frac{x^7}{7!} + 30310\frac{x^8}{8!} + \cdots.$$

Hence the numbers of such graphs on $0, 1, \ldots, 8$ vertices are $0, 0, 1, 4, 15, 72$, $435, 3300, 30310$.

8. One finds $\begin{bmatrix} n \\ k \end{bmatrix} = \begin{bmatrix} n-1 \\ k-1 \end{bmatrix} + (n-1)\begin{bmatrix} n-1 \\ k \end{bmatrix}$. This can be proved directly by considering separately those permutations of n letters and k cycles in which n is a fixed point (cycle of length 1) and those in which it is not.

9. Here the deck enumerator is

$$\mathcal{D}(x) = \sum_{m \geq 1} \frac{(2m-1)! x^{2m}}{(2m)!} = \log \frac{1}{\sqrt{1-x^2}}.$$

The exponential formula states that the question is answered by $\sinh \mathcal{D}(x)$, which simplifies to

$$\mathcal{H}(x) = \frac{x^2}{2\sqrt{1-x^2}}.$$

The coefficient of $x^n/n!$ is $g(n) = 0$ if n is odd, and

$$g(n) = \frac{n!}{2^{n-1}} \binom{n-2}{\frac{n}{2}-1}$$

if $n \geq 2$ is even.

10. We find that

$$\begin{Bmatrix} n \\ k \end{Bmatrix} = \sum_{r=1}^{k} \frac{(-1)^{k-r} r^{n-1}}{(k-r)!(r-1)!}.$$

12. If $F(n,k)$ is the number of n-permutations whose cycles have lengths $\leq k$, then F has the egf $\exp(x + \cdots + x^k/k)$. But $f(n,k)$ counts those whose longest cycle has length k, so if $k \geq 1$, $f(n,k) = F(n,k) - F(n,k-1)$, and the required egf is

$$e^{x + \cdots + \frac{x^{k-1}}{k-1}} \left(e^{\frac{x^k}{k}} - 1 \right).$$

13.

(a) It is

$$\sum_{i+j+k=n} \frac{t_i g_j g_k}{i! j! k!} = 1 \qquad (n \geq 0).$$

(b) If we multiply through by $n!$ to get

$$\sum_{i+j+k=n} \frac{n!}{i! j! k!} t_i g_j g_k = n! \qquad (n \geq 0).$$

then the multinomial coefficient under the summation sign counts
the ways of choosing an ordered triple (R, S, T) of subsets that partition $[n]$, t_i counts the involutions of i letters, each of which gets
relabeled with the elements of R, g_j counts the 2-regular graphs of
j vertices, each of which gets relabeled with the elements of S, etc.
Finally, the right side $n!$ counts n-permutations.

(c) This elegant solution was found by Mr. Douglas Katzman. Given
the triple (τ, G_1, G_2), we construct the corresponding permutation σ
as follows the cycles of the involution τ, acting on R, become cycles
of σ. For each cycle in the graph G_1, locate the smallest numbered
vertex v in the cycle. Choose that one of the two possible ways of
orienting the cycle which carries v to the larger numbered vertex
of its two neighbors. Conversely, in G_2 select the orientation that
carries the smallest numbered vertex of each cycle into the smaller
of its two neighbors.

14.

(a) From the defining equation

$$e^{y\mathcal{D}(x)} = \sum_n \frac{\phi_n(y)}{n!} x^n,$$

we see that each application of the operator D_y multiplies the left
side by another $\mathcal{D}(x)$, so the application of some function $f(D_y)$ will
multiply it by $f(\mathcal{D}(x))$. If we choose f to be the inverse function
$\mathcal{D}^{(-1)}(D_y)$, then we will multiply the left side by $\mathcal{D}^{(-1)}(\mathcal{D}(x))$, that
is, by x. If we multiply the right side of the defining equation by x,
we see that it becomes the egf of $\{n\phi_{n-1}(y)\}$, as claimed.

(b) In this family,

$$e^{y\mathcal{D}(x)} = \frac{1}{(1-x)^y} = \sum_n \binom{-y}{n}(-1)^n x^n$$
$$= \sum_n \frac{y(y+1)\cdots(y+n-1)}{n!} x^n.$$

Thus $\phi_n(y)$ is the 'rising factorial' $y(y+1)\cdots(y+n-1)$. To
check the identity, we have first that the deck enumerator is $\mathcal{D}(x) = -\log(1-x)$. Hence $\mathcal{D}^{(-1)}(x) = 1 - e^{-x}$. Therefore

$$\mathcal{D}^{(-1)}(D_y)\phi_n(y) = \{1 - e^{-D_y}\}\phi_n(y).$$

But Taylor's theorem from differential calculus is identical with the assertion that $(e^D)f(y) = f(y + 1)$ (!!check this!!). Hence

$$(1 - e^{-D_y})\phi_n(y) = \phi_n(y) - \phi_n(y - 1)$$
$$= \{y \cdots (y + n - 1)\} - \{(y - 1) \cdots (y + n - 2)\}$$
$$= ny(y + 1) \cdots (y + n - 2)$$
$$= n\phi_{n-1}(y),$$

as required.

15. The result claimed is certainly true if there is only one card in the deck. Then, by the merge, trickle, and flood argument, it is true in general. Part (b) is immediate. For part (c), insert the factors into the product, and outside the product write the reciprocals of all of those factors, to get

$$p(x) = \prod_{k=1}^{\infty} e^{\frac{x^k}{k}} \prod_{k=1}^{\infty} (1 + \frac{x^k}{k}) e^{-\frac{x^k}{k}}$$
$$= \frac{1}{(1 - x)} \prod_{k=1}^{\infty} (1 + \frac{x^k}{k}) e^{-\frac{x^k}{k}}.$$

Now as $x \to 1^-$, the infinite product approaches a certain universal constant, viz.

$$C = \prod_{k=1}^{\infty} (1 + \frac{1}{k}) e^{-\frac{1}{k}},$$

hence $p(x) \sim C/(1 - x)$. The constant is in fact $e^{-\gamma}$, where γ is Euler's constant.

16. Here c_n is the coefficient of $x^n/n!$ in

$$(1 + x)^{x+1} = (1 + x)^x (1 + x) = (1 + x) \sum_k \binom{x}{k} x^k$$

$$= (1 + x) \sum_k \frac{x(x - 1) \cdots (x - k + 1)}{k!} x^k$$

$$= (1 + x) \sum_k \frac{x^k}{k!} \sum_r (-1)^r \begin{bmatrix} k \\ r \end{bmatrix} x^r.$$

Thus

$$\frac{c_n}{n} = (n - 1)! \left\{ \sum_k \frac{(-1)^{n-k}}{k!} \left(\begin{bmatrix} k \\ n - k \end{bmatrix} - \begin{bmatrix} k \\ n - k - 1 \end{bmatrix} \right) \right\}.$$

But in the sum the terms all vanish for $k \geq n$, hence the right side is an integer.

18. The number of cards in the jth deck is 1 for $j = 1, 2$ and is 2 for $j \geq 3$. Hence by (3.14.6), the hand enumerator is

$$\frac{1}{1-x} \frac{1}{(1-x^2)} \prod_{j \geq 3} \frac{1}{(1-x^j)^2} = \frac{P(x)^2}{(1-x)(1-x^2)}$$

where $P(x)$ is Euler's generating function (3.16.3) for $\{p(n)\}$.

19. In such a tree there is a rooted tree of j vertices attached to one of the edges incident at the root, and a rooted tree of $n - 1 - j$ vertices attached to the other edge at the root. Further, the full tree is completely determined by this unordered pair of trees, and so the number a_n of such full trees is equal to the number of unordered pairs of rooted trees, the total number of whose vertices is $n - 1$, i.e.,

$$a_n = \frac{1}{2} \sum_j t_j t_{n-1-j}$$

if $n - 1$ is odd, for then every unordered pair is counted twice by the sum. If $n - 1$ is even then we need to consider the number of ways that the two subtrees at the root can be of the same size $(n - 1)/2$. The number of unordered pairs of not necessarily distinct objects that can be chosen from a set of a different objects if $\binom{a+1}{2}$. Thus in this case the formula above needs an extra term $t_{(n-1)/2}/2$ added to it, which is equivalent to the result stated.

20. It is 29. 29 cannot be of the form stated, for otherwise we could subtract some multiple of 15 from it to find a nonnegative number of the form $6x + 10y$. But 29 is not of that form since it is odd, and 14 isn't either. Next, if n is any integer that is representable then so is $n + 6$, so to see that every integer larger than 29 is so representable it is enough to observe that $30 = 6 \cdot 5$, $31 = 6 + 10 + 15$, $32 = 6 \cdot 2 + 10 \cdot 2$, $33 = 6 \cdot 3 + 15 \cdot 1$, $34 = 6 \cdot 4 + 10 \cdot 1$, and $35 = 10 \cdot 2 + 15 \cdot 1$.

21. If $f(n)$ is that number then

$$\sum_{n \geq 0} f(n) x^n = \frac{1}{(1-x)(1-x^2)(1-x^3)} = \frac{1}{6(1-x)^3} + \frac{1}{4(1-x)^2}$$
$$+ \frac{17}{72(1-x)} + \frac{1}{8(1+x)} + \frac{1}{9(1-wx)} + \frac{1}{9(1-\bar{w}x)}.$$

If we expand each of the fractions on the right we find the formula

$$f(n) = \frac{1}{6}\binom{n+2}{2} + \frac{1}{4}(n+1) + \frac{17}{72} + \frac{(-1)^n}{8} + \frac{2}{9}\cos\left(\frac{2n\pi}{3}\right)$$

which can be rewritten as

$$f(n) = \frac{(n+3)^2}{12} + \frac{-7 + 9(-1)^n + 16\cos(2n\pi/3)}{72}.$$

The second fraction cannot exceed $32/72 < 1/2$ in absolute value, so $f(n)$ is the unique integer whose distance from $(n+3)^2/12$ is less than $1/2$, as required.

22. For a given $a_1, a_2, \cdots$, we put $n = a_1 + 2a_2 + \cdots$, and we can then construct all possible hands of the desired type by choosing and labeling cards from the given decks as follows.

Make an ordered selection of a_1 cards of size 1 chosen independently from the d_1 cards of size 1 that are available in deck 1. Then make an ordered selection of a_2 cards of size 2 from the d_2 cards of that size that are available in deck 2, etc. The number of ways in which this can be done is $d_1^{a_1} d_2^{a_2} \cdots$.

Next, for the a_1 chosen cards of size 1, choose the 1 label that will appear on each card, which can be done in $n!/(n-a_1)!$ ways, but since the order of these cards in the hand is immaterial, this labeling can be done in only $n!/(a_1!(n-a_1)!)$ ways.

Then, for the a_2 chosen cards of size 2, choose the unordered pairs of labels that will appear on each card. This can be done in

$$\binom{n-a_1}{2}\binom{n-a_1-2}{2}\cdots\binom{n-a_1-2a_2+2}{2} = \frac{(n-a_1)!}{(n-a_1-2a_2)!2!^{a_2}}$$

ways (we need only the unordered pairs because the chosen cards have place-holders on them that tell us in what sequence to place the chosen label set on the card). Finally, since the order of the cards of size 2 in the hand is immaterial, there are only

$$\frac{(n-a_1)!}{(n-a_1-2a_2)!2!^{a_2}a_2!}$$

different ways to do this.

In general, for the a_j chosen cards of size j, we can choose the sequence of sets of j labels that will appear on each card in exactly

$$\frac{(n-a_1-2a_2-\cdots-(j-1)a_{j-1})!}{(n-a_1-2a_2-\cdots-ja_j)!j!^{a_j}a_j!}$$

different ways. If we multiply all of these together, for all $j \geq 1$, we find that the number of hands of the desired specification is

$$\frac{n!d_1^{a_1}d_2^{a_2}\cdots}{1!^{a_1}2!^{a_2}\cdots a_1!a_2!\cdots}.$$

But this is exactly the coefficient of $t^n x_1^{a_1} x_2^{a_2}\cdots/n!$ in the expansion shown in the statement of the problem.

For part (b), in the family of set partitions we have all $d_j = 1$ for $j \geq 1$. Use the result of part (a), with $x_3 = x_4 = \cdots = 1$, since we don't care about classes of size greater than 2, to obtain the joint distribution of classes of sizes 1 and 2 in the form stated.

Answers to problems for chapter 4

1. We have $p_n = (1-p)^{n-1}p$ for $n \geq 1$, hence $\{p_n\}$ has the opsgf $P(x) = px/(1-(1-p)x)$. The mean is $P'(1) = 1/p$, and from (4.1.3) the variance is

$$\sigma^2 = (\log P)' + (\log P)''\big|_{x=1} = \frac{(1-p)}{p^2}.$$

2.

(a) Consider a sequence of n trials that yields a complete collection for the first time at the nth trial. From that sequence we will construct an ordered partition of the set $[n-1]$ into $d-1$ classes, as follows: if the ith photo was chosen at the jth trial ($1 \leq i \leq d$, $1 \leq j \leq n-1$), then put j into the ith class of the partition. Note that $d-1$ of the classes are nonempty. Conversely, from such an ordered partition of $[n-1]$ we can construct exactly d collecting sequences, one for each choice of the coupon that wasn't collected in the first $n-1$ trials. There are $(d-1)!\{^{n-1}_{d-1}\}$ ordered partitions of $[n-1]$ into $d-1$ classes, so there are $d!\{^{n-1}_{d-1}\}$ sequences of trials that obtain a complete collection precisely at the nth trial. There are d^n unrestricted sequences of n trials, so the probability of the event described is as shown.

(b) By (1.6.5),

$$p(x) = (d-1)!x \sum_n \left\{^n_d\right\}\left(\frac{x}{d}\right)^n = \frac{(d-1)!x^d}{(d-x)\cdots(d-(d-1)x)}.$$

(c) $p'(1) = d(1 + \frac{1}{2} + \cdots + \frac{1}{d})$

(d) From (4.1.3),

$$\sigma^2 = d^2 \sum_{i=1}^{d} \frac{1}{i^2} - d\left(1 + \frac{1}{2} + \cdots + \frac{1}{d}\right).$$

(e) About 29 boxes of cereal, with a standard deviation of about 11 boxes.

3. For part (a), the probability $p(j, v_1, T)$ has two components. First, with probability $d_1/(d_1 + 1)$, the walk begins with a step to another vertex of T_1. In that case the probability of a first return after j steps is the same as it was in T_1, which gives a contribution of

$$\frac{d_1}{d_1 + 1} p(j; v_1; T_1)$$

to the answer.

On the other hand, with probability $1/(1 + d_1)$ the walk begins by using the edge (v_1, v_2). In that case the required probability will be the probability that the walk takes exactly $j - 2$ steps in the tree T_2, finishing at v_2 and then crossing back over the edge (v_1, v_2) to vertex v_1.

Fix $m \geq 0$, and consider the following event: the sequence of vertices that the walk visits after crossing to v_2 contains exactly $m + 1$ appearances of vertex v_2 followed by the return to v_1. Hence the sequence looks like

$$v_2, W_1, v_2, W_2, \ldots, W_m, v_2,$$

where each of the W_i is a sequence of vertices of $T_2 - v_2$. The total number of steps in such a walk is $j_1 + \cdots + j_m$, where the j_i are the numbers of steps between consecutive returns to v_2. We need the probability that $j_1 + j_2 + \cdots + j_m = j - 2$. But that is

$$\sum_{j_1 + \cdots + j_m = j-2} p(j_1; v_2; T_2) p(j_2; v_2; T_2) \cdots p(j_m; v_2; T_2) \left(\frac{d_2}{d_2 + 1}\right)^m \left(\frac{1}{d_2 + 1}\right)$$

$$= \frac{1}{d_2 + 1} \left(\frac{d_2}{d_2 + 1}\right)^m [x^{j-2}] F_2(x; v_2; T_2)^m.$$

If we put it all together, we find that $p(j; v_1; T)$ is

$$\frac{d_1}{d_1 + 1} p(j; v_1; T_1) + \sum_{m \geq 0} \frac{\left(\frac{d_2}{d_2+1}\right)^m}{(d_1 + 1)(d_2 + 1)} [x^{j-2}] F_2(x; v_2)^m$$

$$= \frac{d_1}{d_1 + 1} p(j; v_1; T_1) + \frac{1}{d_1 + 1} [x^{j-2}] \frac{1}{d_2 + 1 - d_2 F_2(x; v_2)}.$$

Finally, if we multiply by x^j and sum over j, we obtain the result stated.

In part (d) one has $P_n(x) = x^2/(2 - P_{n-1}(x))$ for $n \geq 2$, with $P_1 = 1$. If one assumes $P_n(x) = A_n(x)/B_n(x)$, then $A_n = x^2 B_{n-1}$ and $B_n = 2B_{n-1} - x^2 B_{n-2}$. This leads to the result that

$$P_n(x) = x^2 \left(\frac{r_+^{n-2} + r_-^{n-2}}{r_+^{n-1} + r_-^{n-1}}\right) \quad (n \geq 2)$$

where $r_\pm = 1 \pm \sqrt{1 - x^2}$.

4. The sequence $\{e_{\leq m}\}$ is obviously generated by

$$\frac{E(x)}{1-x} = \frac{N(x-1)}{1-x}.$$

Since $e_{\geq m} = N(0) - e_{\leq m-1}$, it has the gf

$$\frac{N(0)}{1-x} - x\frac{E(x)}{1-x} = \frac{N(0) - xN(x-1)}{1-x}.$$

5. The board consists of only the diagonal cells of a full $n \times n$ board. To put k nonattacking rooks on this board we can choose any k of the n cells on the board, so $r_k = \binom{n}{k}$. Then (4.2.17) with $j = 0$ gives

$$\sum_k (n-k)! \binom{n}{k}(-1)^k = n!\sum_{k=0}^n \frac{(-1)^k}{k!}$$

for the answer, in agreement with (4.2.10).

6. We have

$$\begin{aligned}
\sum_{m\geq 0} \alpha_m x^m &= \sum_{m\geq 0} x^m \sum_{r\geq m}(-1)^{r-m}N_r \\
&= \sum_{r\geq 0}(-1)^r N_r \sum_{0\leq m\leq r}(-1)^m x^m \\
&= \sum_{r\geq 0}(-1)^r N_r \left\{\frac{1 + (-1)^r x^{r+1}}{1+x}\right\} \\
&= \frac{1}{1+x}\left\{e_0 + \sum_{r\geq 0} N_r x^{r+1}\right\} \\
&= \frac{e_0 + xN(x)}{1+x} = \frac{e_0 + xE(1+x)}{1+x} \\
&= \frac{e_0 + x\{e_0 + e_1(1+x) + \cdots\}}{1+x}.
\end{aligned}$$

Problem 7 is similar.

8. The sum is

$$1 + x + \binom{n}{1}(x^2 + x^3) + \binom{n}{2}(x^4 + x^5) + \cdots$$

$$= (1+x)\left(1 + \binom{n}{1}x^2 + \binom{n}{2}x^4 + \cdots\right)$$

$$= (1+x)(1+x^2)^n.$$

If $f_m(y)$ denotes the sum in question, then Snake Oil finds that

$$\sum_m f_m(y)x^m = (1+x)(1+xy+x^2)^n.$$

See what happens if you try to extend this to the sum that results from replacing '$\lfloor r/2 \rfloor$' by '$\lfloor r/3 \rfloor$' in the sum to be found.

9. The 'objects' Ω are the λ^n possible ways of assigning colors to the vertices of G. For each edge e of the graph G there is a property $P(e)$; a coloring has property $P(e)$ if the two endpoints of edge e have the same color. We seek the number of objects that have exactly 0 properties.

Now consider $N(\supseteq S)$. For a given set S of edges, this is the number of colorings such that at least all of the edges in S are badly colored, i.e., have both endpoints the same color. Think of the graph G_S whose vertices are *all* n of the vertices of the graph G, together with just the edges in S. If all of the edges in S are badly colored, and if C is one of the connected components of G_S, then every vertex in C must have the same color. So the number of ways of assigning colors to the vertices of G such that the edges of S are badly colored is $N(\supseteq S) = \lambda^{\kappa(S)}$, where $\kappa(S)$ is the number of connected components of the graph G_S. Hence

$$P(\lambda; x; G) = \sum_r N_r (x-1)^r,$$

where

$$N_r = \sum_{|S|=r} \lambda^{\kappa(S)}.$$

10. There are k^n possible words, and we take these to be our set of objects Ω. A word has property i if the substring w occurs in the word, beginning in position i of the word. Let S be a given subset of properties, i.e., a set of places where the substring w is to begin. We seek $N(\supseteq S)$, which in this case is the number of words of n letters, chosen from an alphabet of k letters, that have the substring w beginning in all of the positions indicated by S, and maybe elsewhere too. But there are no such words if two of the elements of S differ by $< m$, for then two occurrences of w would overlap, contrary to the hypothesis that they cannot do so.

Hence we suppose that no two elements of S differ by $< m$. Then rm of the characters in the word are specified to be occurrences of w, where $r = |S|$. That leaves $n - rm$ characters to be specified, and that can be done in $N(\supseteq S) = k^{n-rm}$ ways. Hence N_r is k^{n-rm} times the number of subsets S of r elements of $[n-m+1]$ that have no two entries that differ by $< m$. But how many such subsets are there?

Consider a subset S of r elements of $[q]$, no two of whose entries differ by $< m$. If we delete the elements of S from $[q]$, the remaining $q-r$ integers are broken into $r+1$ intervals of consecutive integers whose lengths are $t_0, t_1, \ldots, t_r$, say, where each $t_i \geq m-1$ for $1 \leq i \leq r-1$. The number of ways to choose such integers $t_0, \ldots, t_r$ is clearly

$$[x^{q-r}]\left\{\frac{1}{1-x}\right\}\left\{\frac{x^{m-1}}{1-x}\right\}^{r-1}\left\{\frac{1}{1-x}\right\} = [x^{q-r}]\left\{\frac{x^{(m-1)(r-1)}}{(1-x)^{r+1}}\right\}$$

$$= [x^{q-r-(m-1)(r-1)}]\frac{1}{(1-x)^{r+1}}$$

$$= \binom{q-(m-1)(r-1)}{r}.$$

Thus, since $q = n - m + 1$, $N_r = k^{n-rm}\binom{n-mr}{r}$, and the number of w-free words is

$$\sum_r (-1)^r \binom{n-mr}{r} k^{n-rm}.$$

The Snake Oil method tells us that the answer is also the coefficient of x^n in

$$\frac{1}{1-kx+kx^{m+1}}.$$

In turn this suggests that it might have been easier to do this problem by finding a recurrence relation that is satisfied by the answer, instead of by using the sieve method, but we wanted to show you another example of the sieve method in which the $N(\supseteq S)$'s do not depend only on the cardinality of the set S.

13. This is an example where the Snake Oil method doesn't work immediately because the free parameter n appears too often in the summand. As in example 9, the thing to do is to generalize the problem, in this case to the sum

$$\sum_k (-1)^k \binom{n}{k}\binom{n}{n-m+k}.$$

The latter responds nicely to Snake Oil, after multiplying by x^m etc. Then set $m = n$.

17. We find in part (a) that

$$\frac{\partial}{\partial x}\int_{-B}^{A} F(x,y)dy = \int_{-B}^{A}\frac{\partial F}{\partial x}dy$$

$$= \int_{-B}^{A}\frac{\partial G}{\partial y}dy$$

$$= G(x,A) - G(x,-B) \to 0 \qquad (A, B \to \infty).$$

18.

(a) Since the sum of the d's is $2n-2$, their average is $2-2/n$ which is less than 2, so at least one of the d's must be 1. We can suppose w.l.o.g. that $d_1 = 1$. Then, in every tree whose degree sequence is $\Delta = (d_1, \ldots, d_n)$, vertex 1 is connected to exactly one other vertex. There is an obvious 1-1 correspondence between the trees of degree sequence Δ in which vertex 1 is adjacent to vertex j, for some fixed $j \geq 2$, and the trees of $n-1$ vertices $2, 3, \ldots, n$, in which the vertex degrees are $(d_2, \ldots, d_{j-1}, d_j - 1, d_{j+1}, \ldots, d_n)$. By induction on n, then, the number whose degree sequence is Δ is

$$\sum_{j=2}^{n} \frac{(n-3)!}{(d_2-1)!\cdots(d_{j-1}-1)!(d_j-2)!\cdots(d_n-1)!}$$

$$= \sum_{j=2}^{n} \frac{(n-3)!(d_j-1)}{(d_2-1)!\cdots(d_j-1)!\cdots(d_n-1)!}$$

$$= ((2n-3)-(n-1))\frac{(n-3)!}{(d_2-1)!\cdots(d_n-1)!}$$

$$= \frac{(n-2)!}{(d_1-1)!\cdots(d_n-1)!}$$

as required.

(b) By the multinomial theorem (see exercise 20 of chapter 2),

$$F_n(x_1,\ldots,x_n) = (x_1 x_2 \cdots x_n)(x_1 + \cdots + x_n)^{n-2}.$$

(d) Let a tree T have property i if vertex i is an endpoint. If $S \subseteq [n]$ then the number of trees of n vertices whose set of properties contains S is

$$N(\supseteq S) = (n-|S|)^{n-|S|-2}(n-|S|)^{|S|} = (n-|S|)^{n-2},$$

since the first factor is the number of trees of $n-|S|$ vertices and the second factor is the number of ways we can attach the $|S|$ endpoints to such a tree. The result now follows from the sieve.

(e) In the sieve method, the average number of properties that an object has is always N_1/N, which in this case is

$$\frac{(n-1)^{n-2}n}{n^{n-2}} = n(1-\frac{1}{n})^{n-2} \sim \frac{n}{e}.$$

19.

(a) Evidently we have for all T

$$N(\subseteq S) = \sum_{V \subseteq S} N(= V),$$

by definition. Now substitute this for $N(\subseteq S)$) under the summation sign in the expression given on the right side of the statement
of the problem, interchange the order of summation and verify the
resulting identity.

(b) It is

$$\sum_{n\geq 0}\frac{h_n(S)}{n!}x^n = \prod_{s\in S}\exp\frac{d_s}{s!}x^s.$$

Answers to problems for chapter 5

1. Let $t = 1/A$, $\phi(u) = 1 + u^L$, and $f(u) = u$. Then the equation $u = t\phi(u)$
that is treated by the LIF becomes the present equation. The result follows
after a small calculation involving the binomial theorem.

3. In the LIF, choose the function $f(u)$ that satisfies $f'(u) = 1/\phi(u)$. Then

$$\frac{1}{n}[u^{n-1}]\left\{f'(u)\phi(u)^n\right\} = \frac{1}{u}[u^{n-1}]\phi(u)^{n-1}.$$

On the other hand, if we write $z(t) = f(u(t))$, then

$$[t^n]f(u(t)) = [t^n]z(t) = \frac{1}{n}[t^{n-1}]z'(t) = \frac{1}{n}[t^{n-1}]\frac{tu'(t)}{u(t)}.$$

For the last equality of the problem, differentiate $u = t\phi(u)$ with respect to
t.

4.

(a) Put $\phi(u) = 1 + u + u^2$ in the result of the previous problem to find
that $\gamma_n = [t^n]\frac{1}{1-t(1+2u)}$, where $u = u(t)$ satisfies $u = t(1 + u + u^2)$.
By solving the quadratic equation for u and substituting, we find
the result stated.

(b) Let $x = i/\sqrt{3}$ in problem 3.

5.

(a) Clearly $S_p(n) = [x^n]\left\{(1 + x)^{pn}/(1 - x)\right\}$.

(b) Take $\phi(u) = (1 + u)^p$ and $f'(u) = 1/((1 - u)(1 + u)^p)$ in the LIF, and find

$$\frac{S_p(n)}{n+1} = \frac{1}{n+1}[u^n]\frac{(1+u)^{p(n+1)}}{(1-u)(1+u)^p}$$

$$= [t^{n+1}]f(u(t))$$

$$= \frac{1}{n+1}[t^n]\{f'(u(t))u'(t)\}$$

$$= \frac{1}{n+1}[t^n]\frac{u'(t)}{(1-u)(1+u)^p}$$

$$= \frac{1}{n+1}[t^n]\frac{tu'(t)}{(1-u(t))u(t)}.$$

Since $u = t(1+u)^p$, we find $u' = \frac{u(1+u)}{t(1-(p-1)u)}$, and substitution leads to the result stated.

(c) Equate coefficients of x^n on both sides of the result of part (b).

6. Let $x^{n_1}, \ldots, x^{n_k}$ be the powers of x whose coefficients in $P(x)$ are strictly positive. Then, by Schur's theorem 3.15.2, what is needed is that

$$gcd(n_1, \ldots, n_k) = 1.$$

7.

(a) It is

$$(1+n^a)^n \sim n^{na}\exp\left(n^{1-a} - n^{1-2a}/2 + \cdots\right),$$

where the argument of the exponential terminates after the last positive exponent of n is reached.

(b) As above without the factor n^{na}.

(c) It is ~ 1.

8.

(a) It is admissible because, by Schur's theorem 3.15.2, $e^{z+z^2/2+z^3/3}$ has positive coefficients from some point on.

9. Take $f(u) = (1+u)^k$ and $\phi(u) = (1+u)^2$ in the LIF.

References

An excellent general reference on generating functions is [Co], which contains a wealth of beautiful examples. The volume [GJ] is highly recommended to those who wish to study deeper and more varied uses of generating functions. For excellent surveys of combinatorial asymptotics, see [Be], and [Od]. For other methods that can cope with a wide variety of combinatorial identities see [Eg], [Kn] vol. 1, and [GKP]. For assortments of unapologetically difficult problems in asymptotics with advanced solution techniques, see [Br] and [GK]. [Go] is a catalogue of binomial coefficient identities.

[An]   Andrews, George. *The theory of partitions, Encycl. Math. Appl. vol. 2.* Reading, MA: Addison-Wesley, 1976.

[Bei] Beissinger, Janet. 'Factorization and enumeration of labeled combinatorial objects.' Ph.D. Dissertation, University of Pennsylvania (1981).

[Be]   Bender, Edward A. 'Asymptotic methods in enumeration.' *SIAM Review* **16** (1974), 485-515.

[BG]   Bender, E. A., and Goldman, J. R. 'Enumerative uses of generating functions.' *Indiana Univ. Math. J.* **20** (1971), 753-764.

[Br]   de Bruijn, N. G. *Asymptotic methods in analysis.* North Holland, 1958.

[Co]   Comtet, Louis. *Advanced Combinatorics; The art of finite and infinite expansions.* Boston, MA: D. Reidel Publ. Co., 1974.

[De1] Delest, M. P. 'Generating functions for column-convex polyominoes.' *J. Combinatorial Theory A* **48** (1988), 12-31.

[De2] Delest, M. P. 'Polyominoes and animals: some recent results.' J. Mathematical Chemistry **8** (1991), 3-18.

[DV]   Delest, M. P. and Viennot, G. 'Algebraic languages and polyominoes enumeration.' *Theoretical Computer Science* **34** (1984), 169-206.

[DRS] Doubilet, P., Rota, G. C., and Stanley, R. P. 'On the Foundations of Combinatorial Theory VI: The idea of generating function.' In *Proc. Sixth Berkeley Symposium on Statistics and Probability*, vol. 2 (1972), 267-318.

[Eg]   Egorychev, G.P. 'Integral representation and the computation of combinatorial sums.' *American Mathematical Society Translations* **59**, (1984).

[Fo1] Foata, D. 'La Série Génératrice Exponentielle dans les Problèmes d'Énumération.' Montreal: Presses de l'Université de Montréal, 1971.

[Fo2] Foata, D. 'A combinatorial proof of the Mehler formula.' *J. Comb. Th. Ser. A* **24** (1978), 367-376.

[FS]   Foata, D. and Schützenberger, M. *Théorie Géométrique des Polynômes Eulériens,* Lecture Notes in Math. No. 138. Berlin: Springer-Verlag, 1970.

[Fr] Fraenkel, Aviezri. 'A characterization of exactly covering congruences.' *Discrete Mathematics* **4** (1973), 359-366.

[GaJ] Garsia, A. and Joni, S. A. 'Composition sequences.' *Commun. in Algebra* **8** (1980), 1195-1266.

[Gos] Gosper, R. William, Jr. 'Decision procedures for indefinite hypergeometric summation.' *Proc. Nat. Acad. Sci. U.S.A.* **75** (1978), 40-42.

[Go] Gould, Henry W. *Combinatorial identities.* Morgantown, WV, 1972.

[GJ] Goulden, I. P. and Jackson, D. M. *Combinatorial enumeration.* New York: John Wiley and Sons, 1983.

[GKP] Graham, Ronald L., Knuth, Donald. E., and Patashnik, Oren. *Concrete Mathematics.* Reading, MA: Addison-Wesley, 1989.

[GK] Greene, Daniel H. and Knuth, Donald E. *Mathematics for the analysis of algorithms.* Boston: Birkhäuser, 1982.

[Ha] Hansen, E. R. *A table of series and products*, Prentice-Hall, 1975.

[HRS] Harary, F., Robinson, R. W., and Schwenk, A. J. 'A twenty step algorithm for determining the asymptotic number of trees of various species.' *J. Austral Math. Soc. Ser. A* **20** (1975), 483-503.

[Ha] Hayman, Walter. 'A generalisation of Stirling's formula.' *Journal für die reine und angewandte Mathematik* **196** (1956), 67-95.

[Jo] Joyal, A. 'Une théorie combinatoire des séries formelles.' *Adv. Math.* **42** (1981), 1-82.

[Kl] Klarner, D. 'Some results concerning polyominoes.' *Fibonacci Quart.* **3** (1965), 9-20.

[Kn] Knuth, Donald E. *The Art of Computer Programming, vol. 1: Fundamental Algorithms*, 1968 (2nd ed. 1973); *vol. 2: Seminumerical Algorithms*, 1969 (2nd ed. 1981); *vol. 3: Sorting and Searching*, 1973. Reading, MA: Addison-Wesley.

[KnW] Knuth, Donald E., and Wilf, Herbert S. 'A short proof of Darboux's lemma.' *Applied Mathematics Letters* **2** (1989), 139-140.

[Ko] Koepf, Wolfram. 'Power series in computer algebra.' *J. Symb. Comp.*, to appear.

[MP] Moon, J. W. and Pullman, N. J. 'The number of triangles in a triangular lattice.' *Delta* **3** (1973), 28-31.

[NW] Nijenhuis, Albert, and Wilf, Herbert S. 'Representations of integers by linear forms in nonnegative integers.' *J. Number Theory* **4** (1970), 98-106.

[Od] Odlyzko, A. M., Asymptotic enumeration methods, to appear.

[Pe] Petkovšek, Marko. *Finding closed-form solutions of difference equations by symbolic methods*, Ph.D. Dissertation, School of Computer Science, Carnegie Mellon University, CMU-CS-91-103, 1991.

[Po] Porubský, Štefan. *Results and problems on covering systems of residue classes.* Mitteilungen Mathem. Seminar Giessen **150**, Selbstverlag Math. Inst., Giessen, !981.

[Ra] Rademacher, Hans, *Lectures on elementary number theory*, Blaisdell, 1964.

[RU] Riddell, R. J., and Uhlenbeck, G. E. 'On the theory of the virial development of the equation of state of monatomic gases.' *J. Chem. Phys.* **21** (1953), 2056-2064.

[RM] Rota, Gian-Carlo, and Mullin, Ronald. 'On the foundations of combinatorial theory, III.' In *Graph Theory and its Applications*, Reading, MA: Academic Press, 1970, 167-213.

[Ro] Roy, Ranjan. 'Binomial identities and hypergeometric series.' *American Mathematical Monthly* **94** (1987), 36-46.

[Sc] Schützenberger, M. P. 'Context-free languages and pushdown automata.' *Information and Control* **6** (1963), 246-264.

[St1] Stanley, Richard P. *Enumerative combinatorics*. Monterey, CA: Wadsworth, 1986.

[St2] Stanley, Richard P. 'Generating functions.' In *MAA Studies in Combinatorics*, Washington, DC: Mathematical Association of America, 1978.

[St3] Stanley, Richard P. 'Exponential structures.' *Studies in Appl. Math.* **59** (1978), 78-82.

[Sz] Szegö, Gabor. 'Orthogonal polynomials.' *American Mathematical Society Colloquium Series Publication*, 1967.

[Wi1] Wilf, Herbert S. 'Mathematics for the Physical Sciences.' New York: John Wiley and Sons, 1962; reprinted by Dover Publications, 1978.

[Wi2] Wilf, Herbert S. *Algorithms and complexity*. Englewood Cliffs, NJ: Prentice Hall, 1986.

[Wi3] Wilf, Herbert S. 'Three problems in combinatorial asymptotics.' *J. Combinatorial Theory* **35** (1983), 199-207.

[WZ1] Wilf, Herbert S., and Zeilberger, Doron. 'Rational functions certify combinatorial identities.' *J. Amer. Math. Soc.* **3** (1990), 147-158.

[WZ2] Wilf, Herbert S., and Zeilberger, Doron. 'An algorithmic proof theory for hypergeometric (ordinary and 'q') multisum/integral identities.' *Inventiones Mathematicæ*, **108** (1992), 575-633.

[Zn] Znám, Š., 'A survey of covering systems of congruences.' *Acta Math. Univ. Comenian.*, **40-41** *(1982), 59-72.*

Index

algebraic singularities, 177
Andrews, G., 100
animals, 150
asymptotic methods, 167*ff.*
averages, 2, 108

Beissinger, J., 103
Bell numbers, 20–23, 42, 175, 193
Bender, E., 103
Bernoulli numbers, 56, 68, 129, 166
    polynomials, 166
Bessel functions, 165
binomial coefficients, 11
binomial type, 91
bipartite graphs, 87
Bonferroni's inequalities, 158
boundary value problem, 10

calculus of opsgf's, 34*ff.*
Catalan numbers, 44
Cauchy formula, 50
    inequality, 50
Cayley's formula, 163
cereal, boxes of, 157
chessboards, 116, 158
circular arrangements, 26 (ex. 14)
congruences, 140
connected graphs, 73*ff.*, 86
conventions, summation, 13
convex polymino, 151
coupon collector's problem, 157
cycle index, 141
cycle type, 142
cycles of permutations, 81, 109, 176
cyclotomic polynomials

Darboux lemma, 178
derangements, 45, 69 (ex. 7), 144
derivative, of formal power series, 32
Dirichlet series, 56*ff.*
divisor function, 59
Dixon's identity, 136

entire function, 181
Euler, L., 102, 110

Euler's function, 66 (ex. 14)
exact covering sequences, 154
exponential formula, 81
exponential generating function, 21,
    39–45

Fibonacci numbers, 1, 8, 9, 15, 34, 38, 67
fixed points, 45, 69, 113, 144
Foata, D., 103
forest, 90
fountains, 38–39

Garsia, A., 103
geometric series, 4, 6, 16
Goldman, J., 103
Gosper, R. W., Jr., 135
Graham, R. L., 119

Hardy, G., 100
harmonic numbers, 38, 56, 109, 176
Hayman, W. K., 50, 181*ff.*
    admissible, 184
HC-polymino, 151
hypergeometric functions, 128

inverse, of a series, 31
inversion, of a permutation, 17 (ex. 17),
    139
    formulas, 127, 169
involutions, 84, 104 (ex. 5), 184

Joni, S., 103
Joyal, A., 103

Klarner, D., 153
Knuth, D. E., 119, 159

Lagrange inversion formula, 91, 167, 188
Lambert series, 70 (ex. 31)
Legendre polynomials, 188
limit superior, 46
logarithmic concavity, 136

Macsyma, 135
Maple, 192